Universitext

T0202996

Jiří Matoušek
Bernd Gärtner

Understanding and Using Linear Programming

 Springer

Jiří Matoušek

Department of Applied Mathematics
Charles University
Malostranské nám. 25
118 00 Praha 1, Czech Republic
E-mail: matousek@kam.mff.cuni.cz

Bernd Gärtner

Institute of Theoretical Computer Science
ETH Zurich
CH-8092 Zurich, Switzerland
E-mail: gaertner@inf.ethz.ch

Mathematics Subject Classification (2000): 90C05

Library of Congress Control Number: 2006931795

ISBN-10 3-540-30697-8 Springer Berlin Heidelberg New York
ISBN-13 978-3-540-30697-9 Springer Berlin Heidelberg New York

Springer is a part of Springer Science+Business Media
springer.com
© Springer-Verlag Berlin Heidelberg 2007

Typesetting: by the authors and techbooks using a Springer TeX macro package

Cover design: *design & production* GmbH, Heidelberg

Printed on acid-free paper   SPIN: 11592457   46/techbooks   5 4 3 2 1 0

Preface

This is an introductory textbook of linear programming, written mainly for students of computer science and mathematics. Our guiding phrase is, "what every theoretical computer scientist should know about linear programming."

The book is relatively concise, in order to allow the reader to focus on the basic ideas. For a number of topics commonly appearing in thicker books on the subject, we were seriously tempted to add them to the main text, but we decided to present them only very briefly in a separate glossary. At the same time, we aim at covering the main results with complete proofs and in sufficient detail, in a way ready for presentation in class.

One of the main focuses is applications of linear programming, both in practice and in theory. Linear programming has become an extremely flexible tool in theoretical computer science and in mathematics. While many of the finest modern applications are much too complicated to be included in an introductory text, we hope to communicate some of the flavor (and excitement) of such applications on simpler examples.

We present three main computational methods. The simplex algorithm is first introduced on examples, and then we cover the general theory, putting less emphasis on implementation details. For the ellipsoid method we give the algorithm and the main claims required for its analysis, omitting some technical details. From the vast family of interior point methods, we concentrate on one of the most efficient versions known, the primal–dual central path method, and again we do not present the technical machinery in full. Rigorous mathematical statements are clearly distinguished from informal explanations in such parts.

The only real prerequisite to this book is undergraduate linear algebra. We summarize the required notions and results in an appendix. Some of the examples also use rudimentary graph-theoretic terminology, and at several places we refer to notions and facts from calculus; all of these should be a part of standard undergraduate curricula.

Errors. If you find errors in the book, especially serious ones, we would appreciate it if you would let us know (email: matousek@kam.mff.cuni.cz, gaertner@inf.ethz.ch). We plan to post a list of errors at http://www.inf.ethz.ch/personal/gaertner/lpbook.

Acknowledgments. We would like to thank the following people for help, such as reading preliminary versions and giving us invaluable comments: Pierre Dehornoy, David Donoho, Jiří Fiala, Michal Johanis, Volker Kaibel, Edward Kim, Petr Kolman, Jesús de Loera, Nathan Linial, Martin Loebl, Helena Nyklová, Yoshio Okamoto, Jiří Rohn, Leo Rüst, Rahul Savani, Andreas Schulz, Petr Škovroň, Bernhard von Stengel, Tamás Terlaky, Louis Theran, Jiří Tůma, and Uli Wagner. We also thank David Kramer for thoughtful copy-editing.

Prague and Zurich, July 2006   Jiří Matoušek, Bernd Gärtner

Contents

1. What Is It, and What For?

Linear programming, surprisingly, is not directly related to computer programming. The term was introduced in the 1950s when computers were few and mostly top secret, and the word programming was a military term that, at that time, referred to plans or schedules for training, logistical supply, or deployment of men. The word linear suggests that feasible plans are restricted by linear constraints (inequalities), and also that the quality of the plan (e.g., costs or duration) is also measured by a linear function of the considered quantities. In a similar spirit, linear programming soon started to be used for planning all kinds of economic activities, such as transport of raw materials and products among factories, sowing various crop plants, or cutting paper rolls into shorter ones in sizes ordered by customers. The phrase "planning with linear constraints" would perhaps better capture this original meaning of linear programming. However, the term linear programming has been well established for many years, and at the same time, it has acquired a considerably broader meaning: Not only does it play a role only in mathematical economy, it appears frequently in computer science and in many other fields.

1.1 A Linear Program

We begin with a very simple linear programming problem (or **linear program** for short):

Maximize the value $x_1 + x_2$
among all vectors $(x_1, x_2) \in \mathbb{R}^2$
satisfying the constraints
$$x_1 \geq 0$$
$$x_2 \geq 0$$
$$x_2 - x_1 \leq 1$$
$$x_1 + 6x_2 \leq 15$$
$$4x_1 - x_2 \leq 10.$$

For this linear program we can easily draw a picture. The set $\{\mathbf{x} \in \mathbb{R}^2 : x_2 - x_1 \leq 1\}$ is the half-plane lying below the line $x_2 = x_1 + 1$, and similarly,

each of the remaining four inequalities defines a half-plane. The set of all
vectors satisfying the five constraints simultaneously is a convex polygon:

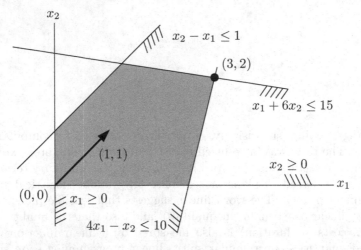

Which point of this polygon maximizes the value of $x_1 + x_2$? The one lying
"farthest in the direction" of the vector $(1,1)$ drawn by the arrow; that is,
the point $(3,2)$. The phrase "farthest in the direction" is in quotation marks
since it is not quite precise. To make it more precise, we consider a line
perpendicular to the arrow, and we think of translating it in the direction of
the arrow. Then we are seeking a point where the moving line intersects our
polygon for the last time. (Let us note that the function $x_1 + x_2$ is constant
on each line perpendicular to the vector $(1,1)$, and as we move the line in
the direction of that vector, the value of the function increases.) See the next
illustration:

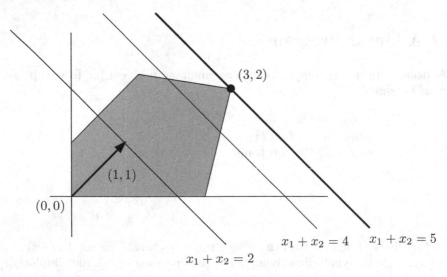

In a general linear program we want to find a vector $\mathbf{x}^* \in \mathbb{R}^n$ maximizing (or minimizing) the value of a given linear function among all vectors $\mathbf{x} \in \mathbb{R}^n$ that satisfy a given system of linear equations and inequalities. The linear function to be maximized, or sometimes minimized, is called the **objective function**. It has the form $\mathbf{c}^T\mathbf{x} = c_1x_1 + \cdots + c_nx_n$, where $\mathbf{c} \in \mathbb{R}^n$ is a given vector.[1]

The linear equations and inequalities in the linear program are called the **constraints**. It is customary to denote the number of constraints by m.

A linear program is often written using matrices and vectors, in a way similar to the notation $A\mathbf{x} = \mathbf{b}$ for a system of linear equations in linear algebra. To make such a notation simpler, we can replace each equation in the linear program by two opposite inequalities. For example, instead of the constraint $x_1 + 3x_2 = 7$ we can put the two constraints $x_1 + 3x_2 \leq 7$ and $x_1 + 3x_2 \geq 7$. Moreover, the direction of the inequalities can be reversed by changing the signs: $x_1 + 3x_2 \geq 7$ is equivalent to $-x_1 - 3x_2 \leq -7$, and thus we can assume that all inequality signs are "$\leq$", say, with all variables appearing on the left-hand side. Finally, minimizing an objective function $\mathbf{c}^T\mathbf{x}$ is equivalent to maximizing $-\mathbf{c}^T\mathbf{x}$, and hence we can always pass to a maximization problem. After such modifications each linear program can be expressed as follows:

Maximize the value of $\qquad\qquad \mathbf{c}^T\mathbf{x}$
among all vectors $\mathbf{x} \in \mathbb{R}^n$ satisfying $\quad A\mathbf{x} \leq \mathbf{b}$,

where A is a given $m \times n$ real matrix and $\mathbf{c} \in \mathbb{R}^n$, $\mathbf{b} \in \mathbb{R}^m$ are given vectors. Here the relation $\leq$ holds for two vectors of equal length if and only if it holds componentwise.

Any vector $\mathbf{x} \in \mathbb{R}^n$ satisfying all constraints of a given linear program is a **feasible solution**. Each $\mathbf{x}^* \in \mathbb{R}^n$ that gives the maximum possible value of $\mathbf{c}^T\mathbf{x}$ among all feasible $\mathbf{x}$ is called an **optimal solution**, or **optimum** for short. In our linear program above we have $n = 2$, $m = 5$, and $\mathbf{c} = (1, 1)$. The only optimal solution is the vector $(3, 2)$, while, for instance, $(2, \frac{3}{2})$ is a feasible solution that is not optimal.

A linear program may in general have a single optimal solution, or infinitely many optimal solutions, or none at all.

We have seen a situation with a single optimal solution in the first example of a linear program. We will present examples of the other possible situations.

---

[1] Here we regard the vector $\mathbf{c}$ as an $n \times 1$ matrix, and so the expression $\mathbf{c}^T\mathbf{x}$ is a product of a $1 \times n$ matrix and an $n \times 1$ matrix. This product, formally speaking, should be a 1×1 matrix, but we regard it as a real number.

Some readers might wonder: If we consider $\mathbf{c}$ a column vector, why, in the example above, don't we write it as a column or as $(1, 1)^T$? For us, a vector is an n-tuple of numbers, and when writing an explicit vector, we separate the numbers by commas, as in $\mathbf{c} = (1, 1)$. Only if a vector appears in a context where one expects a matrix, that is, in a product of matrices, *then* it is regarded as (or "converted to") an $n \times 1$ matrix. (However, sometimes we declare a vector to be a row vector, and then it behaves as a $1 \times n$ matrix.)

If we change the vector **c** in the example to $(\frac{1}{6}, 1)$, all points on the side of the polygon drawn thick in the next picture are optimal solutions:

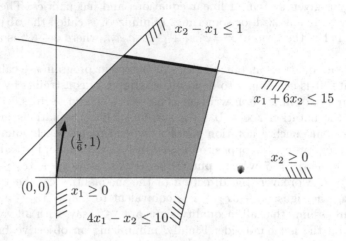

If we reverse the directions of the inequalities in the constraints $x_2 - x_1 \leq 1$ and $4x_1 - x_2 \leq 10$ in our first example, we obtain a linear program that has no feasible solution, and hence no optimal solution either:

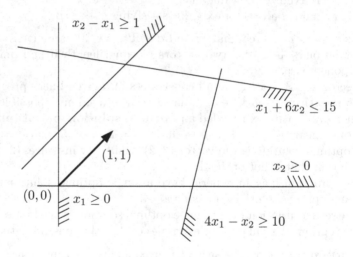

Such a linear program is called **infeasible**.

Finally, an optimal solution need not exist even when there are feasible solutions. This happens when the objective function can attain arbitrarily large values (such a linear program is called **unbounded**). This is the case when we remove the constraints $4x_1 - x_2 \leq 10$ and $x_1 + 6x_2 \leq 15$ from the initial example, as shown in the next picture:

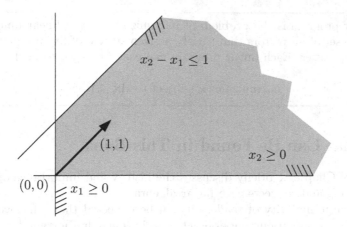

Let us summarize: We have seen that a linear program can have one or infinitely many optimal solutions, but it may also be unbounded or infeasible. Later we will prove that no other situations can occur.

We have solved the initial linear program graphically. It was easy since there are only two variables. However, for a linear program with four variables we won't even be able to make a picture, let alone find an optimal solution graphically. A substantial linear program in practice often has several thousand variables, rather than two or four. A graphical illustration is useful for understanding the notions and procedures of linear programming, but as a computational method it is worthless. Sometimes it may even be misleading, since objects in high dimension may behave in a way quite different from what the intuition gained in the plane or in three-dimensional space suggests.

One of the key pieces of knowledge about linear programming that one should remember forever is this:

> A linear program is efficiently solvable, both in theory and in practice.

- *In practice*, a number of software packages are available. They can handle inputs with several thousands of variables and constraints. Linear programs with a special structure, for example, with a small number of nonzero coefficients in each constraint, can often be managed even with a much larger number of variables and constraints.
- *In theory*, algorithms have been developed that provably solve each linear program in time bounded by a certain polynomial function of the input size. The input size is measured as the total number of bits needed to write down all coefficients in the objective function and in all the constraints.

These two statements summarize the results of long and strenuous research, and efficient methods for linear programming are not simple.

In order that the above piece of knowledge will also make sense forever, one should not forget what a linear program is, so we repeat it once again:

> A linear program is the problem of maximizing a given linear function
> over the set of all vectors that satisfy a given system of linear equations
> and inequalities. Each linear program can easily be transformed to the
> form
>
> $$\text{maximize } \mathbf{c}^T \mathbf{x} \text{ subject to } A\mathbf{x} \le \mathbf{b}.$$

1.2 What Can Be Found in This Book

The rest of Chapter 1 briefly discusses the history and importance of linear programming and connects it to linear algebra.

For a large majority of readers it can be expected that whenever they encounter linear programming in practice or in research, they will be using it as a black box. From this point of view Chapter 2 is crucial, since it describes a number of algorithmic problems that can be solved via linear programming.

The closely related Chapter 3 discusses integer programming, in which one also optimizes a linear function over a set of vectors determined by linear constraints, but moreover, the variables must attain integer values. In this context we will see how linear programming can help in approximate solutions of hard computational problems.

Chapter 4 brings basic theoretical results on linear programming and on the geometric structure of the set of all feasible solutions. Notions introduced there, such as convexity and convex polyhedra, are important in many other branches of mathematics and computer science as well.

Chapter 5 covers the simplex method, which is a fundamental algorithm for linear programming. In full detail it is relatively complicated, and from the contemporary point of view it is not necessarily the central topic in a first course on linear programming. In contrast, some traditional introductions to linear programming are focused almost solely on the simplex method.

In Chapter 6 we will state and prove the duality theorem, which is one of the principal theoretical results in linear programming and an extremely useful tool for proofs.

Chapter 7 deals with two other important algorithmic approaches to linear programming: the ellipsoid method and the interior point method. Both of them are rather intricate and we omit some technical issues.

Chapter 8 collects several slightly more advanced applications of linear programming from various fields, each with motivation and some background material.

Chapter 9 contains remarks on software available for linear programming and on the literature.

Linear algebra is the main mathematical tool throughout the book. The required linear-algebraic notions and results are summarized in an appendix.

The book concludes with a glossary of terms that are common in linear programming but do not appear in the main text. Some of them are listed to ensure that our index can compete with those of thicker books, and others appear as background material for the advanced reader.

Two levels of text. This book should serve mainly as an introductory text for undergraduate and early graduate students, and so we do not want to assume previous knowledge beyond the usual basic undergraduate courses. However, many of the key results in linear programming, which would be a pity to omit, are not easy to prove, and sometimes they use mathematical methods whose knowledge cannot be expected at the undergraduate level. Consequently, the text is divided into two levels. On the basic level we are aiming at full and sufficiently detailed proofs.

> The second, more advanced, and "edifying" level is typographically distinguished like this. In such parts, intended chiefly for mathematically more mature readers, say graduate or PhD students, we include sketches of proofs and somewhat imprecise formulations of more advanced results. Whoever finds these passages incomprehensible may freely ignore them; the basic text should also make sense without them.

1.3 Linear Programming and Linear Algebra

The basics of linear algebra can be regarded as a theory of systems of linear equations. Linear algebra considers many other things as well, but systems of linear equations are surely one of the core subjects. A key algorithm is Gaussian elimination, which efficiently finds a solution of such a system, and even a description of the set of all solutions. Geometrically, the solution set is an affine subspace of $\mathbb{R}^n$, which is an important linear-algebraic notion.[2]

In a similar spirit, the discipline of linear programming can be regarded as a theory of systems of linear *inequalities*.

> In a linear program this is somewhat obscured by the fact that we do not look for an arbitrary solution of the given system of inequalities, but rather a solution maximizing a given objective function. But it can be shown that finding an (arbitrary) feasible solution of a linear program, if one exists, is computationally almost equally difficult as finding an optimal solution. Let us outline how one can gain an optimal solution, provided that feasible solutions can be computed (a different and more elegant way will be described in Section 6.1). If we somehow know in advance that, for instance, the maximum value of the objective function in a given linear program lies between 0 and 100, we can first ask, whether there exists a feasible $\mathbf{x} \in \mathbb{R}^n$ for which the objective

---

[2] An affine subspace is a linear subspace translated by a fixed vector $\mathbf{x} \in \mathbb{R}^n$. For example, every point, every line, and $\mathbb{R}^2$ itself are the affine subspaces of $\mathbb{R}^2$.

function is at least 50. That is, we add to the existing constraints a new constraint requiring that the value of the objective function be at least 50, and we find out whether this auxiliary linear program has a feasible solution. If yes, we will further ask, by the same trick, whether the objective function can be at least 75, and if not, we will check whether it can be at least 25. A reader with computer-science-conditioned reflexes has probably already recognized the strategy of *binary search*, which allows us to quickly localize the maximum value of the objective function with great accuracy.

Geometrically, the set of all solutions of a system of linear inequalities is an intersection of finitely many half-spaces in $\mathbb{R}^n$. Such a set is called a *convex polyhedron*, and familiar examples of convex polyhedra in $\mathbb{R}^3$ are a cube, a rectangular box, a tetrahedron, and a regular dodecahedron. Convex polyhedra are mathematically much more complex objects than vector subspaces or affine subspaces (we will return to this later). So actually, we can be grateful for the objective function in a linear program: It is enough to compute a single point $\mathbf{x}^* \in \mathbb{R}^n$ as a solution and we need not worry about the whole polyhedron.

In linear programming, a role comparable to that of Gaussian elimination in linear algebra is played by the *simplex method*. It is an algorithm for solving linear programs, usually quite efficient, and it also allows one to prove theoretical results.

Let us summarize the analogies between linear algebra and linear programming in tabular form:

	Basic problem	Algorithm	Solution set
Linear algebra	system of linear equations	Gaussian elimination	affine subspace
Linear programming	system of linear inequalities	simplex method	convex polyhedron

1.4 Significance and History of Linear Programming

In a special issue of the journal *Computing in Science & Engineering*, the simplex method was included among "the ten algorithms with the greatest influence on the development and practice of science and engineering in the 20th century."[3] Although some may argue that the simplex method is only

---

[3] The remaining nine algorithms on this list are the Metropolis algorithm for Monte Carlo simulations, the Krylov subspace iteration methods, the decompositional approach to matrix computations, the Fortran optimizing compiler, the QR algorithm for computing eigenvalues, the Quicksort algorithm for sorting, the fast Fourier transform, the detection of integer relations, and the fast multipole method.

number fourteen, say, and although each such evaluation is necessarily subjective, the importance of linear programming can hardly be cast in doubt.

The simplex method was invented and developed by George Dantzig in 1947, based on his work for the U.S. Air Force. Even earlier, in 1939, Leonid Vitalyevich Kantorovich was charged with the reorganization of the timber industry in the U.S.S.R., and as a part of his task he formulated a restricted class of linear programs and a method for their solution. As happens under such regimes, his discoveries went almost unnoticed and nobody continued his work. Kantorovich together with Tjalling Koopmans received the Nobel Prize in Economics in 1975, for pioneering work in resource allocation. Somewhat ironically, Dantzig, whose contribution to linear programming is no doubt much more significant, was never awarded a Nobel Prize.

The discovery of the simplex method had a great impact on both theory and practice in economics. Linear programming was used to allocate resources, plan production, schedule workers, plan investment portfolios, and formulate marketing and military strategies. Even entrepreneurs and managers accustomed to relying on their experience and intuition were impressed when costs were cut by 20%, say, by a mere reorganization according to some mysterious calculation. Especially when such a feat was accomplished by someone who was not really familiar with the company, just on the basis of some numerical data. Suddenly, mathematical methods could no longer be ignored with impunity in a competitive environment.

Linear programming has evolved a great deal since the 1940s, and new types of applications have been found, by far not restricted to mathematical economics.

In theoretical computer science it has become one of the fundamental tools in algorithm design. For a number of computational problems the existence of an efficient (polynomial-time) algorithm was first established by general techniques based on linear programming.

For other problems, known to be computationally difficult (NP-hard, if this term tells the reader anything), finding an exact solution is often hopeless. One looks for approximate algorithms, and linear programming is a key component of the most powerful known methods.

Another surprising application of linear programming is theoretical: the *duality theorem*, which will be explained in Chapter 6, appears in proofs of numerous mathematical statements, most notably in combinatorics, and it provides a unifying abstract view of many seemingly unrelated results. The duality theorem is also significant algorithmically.

We will show examples of methods for constructing algorithms and proofs based on linear programming, but many other results of this kind are too advanced for a short introductory text like ours.

The theory of algorithms for linear programming itself has also grown considerably. As everybody knows, today's computers are many orders of magnitude faster than those of fifty years ago, and so it doesn't sound surprising that much larger linear programs can be solved today. But it may be surprising that this enlargement of manageable problems probably owes more to theoretical progress in algorithms than to faster computers. On the one hand, the implementation of the simplex method has been refined considerably, and on the other hand, new computational methods based on completely different ideas have been developed. This latter development will be described in Chapter 7.

2. Examples

Linear programming is a wonderful tool. But in order to use it, one first has to start suspecting that the considered computational problem might be expressible by a linear program, and then one has to really express it that way. In other words, one has to see linear programming "behind the scenes."

One of the main goals of this book is to help the reader acquire skills in this direction. We believe that this is best done by studying diverse examples and by practice. In this chapter we present several basic cases from the wide spectrum of problems amenable to linear programming methods, and we demonstrate a few tricks for reformulating problems that do not look like linear programs at first sight. Further examples are covered in Chapter 3, and Chapter 8 includes more advanced applications.

Once we have a suitable linear programming formulation (a "model" in the mathematical programming parlance), we can employ general algorithms. From a programmer's point of view this is very convenient, since it suffices to input the appropriate objective function and constraints into general-purpose software.

If efficiency is a concern, this need not be the end of the story. Many problems have special features, and sometimes specialized algorithms are known, or can be constructed, that solve such problems substantially faster than a general approach based on linear programming. For example, the study of network flows, which we consider in Section 2.2, constitutes an extensive subfield of theoretical computer science, and fairly efficient algorithms have been developed. Computing a maximum flow via linear programming is thus not the best approach for large-scale instances.

However, even for problems where linear programming doesn't ultimately yield the most efficient available algorithm, starting with a linear programming formulation makes sense: for fast prototyping, case studies, and deciding whether developing problem-specific software is worth the effort.

2.1 Optimized Diet: Wholesome and Cheap?

> *. . . and when Rabbit said, "Honey or condensed milk with your bread?" he was so excited that he said, "Both," and then, so as not to seem greedy, he added, "But don't bother about the bread, please."*

> A. A. Milne, Winnie the Pooh

The Office of Nutrition Inspection of the EU recently found out that dishes served at the dining and beverage facility "Bullneck's," such as herring, hot dogs, and house-style hamburgers do not comport with the new nutritional regulations, and its report mentioned explicitly the lack of vitamins A and C and dietary fiber. The owner and operator of the aforementioned facility is attempting to rectify these shortcomings by augmenting the menu with vegetable side dishes, which he intends to create from white cabbage, carrots, and a stockpile of pickled cucumbers discovered in the cellar. The following table summarizes the numerical data: the prescribed amount of the vitamins and fiber per dish, their content in the foods, and the unit prices of the foods.[1]

Food	Carrot, Raw	White Cabbage, Raw	Cucumber, Pickled	Required per dish
Vitamin A [mg/kg]	35	0.5	0.5	0.5 mg
Vitamin C [mg/kg]	60	300	10	15 mg
Dietary Fiber [g/kg]	30	20	10	4 g
price [€/kg]	0.75	0.5	0.15*	—

*Residual accounting price of the inventory, most likely unsaleable.

At what minimum additional price per dish can the requirements of the Office of Nutrition Inspection be satisfied? This question can be expressed by the following linear program:

$$
\begin{aligned}
\text{Minimize} \quad & 0.75x_1 + 0.5x_2 + 0.15x_3 \\
\text{subject to} \quad & x_1 \geq 0 \\
& x_2 \geq 0 \\
& x_3 \geq 0 \\
& 35x_1 + 0.5x_2 + 0.5x_3 \geq 0.5 \\
& 60x_1 + 300x_2 + 10x_3 \geq 15 \\
& 30x_1 + 20x_2 + 10x_3 \geq 4.
\end{aligned}
$$

The variable x_1 specifies the amount of carrot (in kg) to be added to each dish, and similarly for x_2 (cabbage) and x_3 (cucumber). The objective function

---

[1] For those interested in healthy diet: The vitamin contents and other data are more or less realistic.

expresses the price of the combination. The amounts of carrot, cabbage, and cucumber are always nonnegative, which is captured by the conditions $x_1 \geq 0$, $x_2 \geq 0$, $x_3 \geq 0$ (if we didn't include them, an optimal solution might perhaps have the amount of carrot, say, negative, by which one would seemingly save money). Finally, the inequalities in the last three lines force the requirements on vitamins A and C and of dietary fiber.

The linear program can be solved by standard methods. The optimal solution yields the price of €0.07 with the following doses: carrot 9.5 g, cabbage 38 g, and pickled cucumber 290 g per dish (all rounded to two significant digits). This probably wouldn't pass another round of inspection. In reality one would have to add further constraints, for example, one on the maximum amount of pickled cucumber.

We have included this example so that our treatment doesn't look too revolutionary. It seems that all introductions to linear programming begin with various dietary problems, most likely because the first large-scale problem on which the simplex method was tested in 1947 was the determination of an adequate diet of least cost. Which foods should be combined and in what amounts so that the required amounts of all essential nutrients are satisfied and the daily ration is the cheapest possible. The linear program had 77 variables and 9 constraints, and its solution by the simplex method using hand-operated desk calculators took approximately 120 man-days.

Later on, when George Dantzig had already gained access to an electronic computer, he tried to optimize his own diet as well. The optimal solution of the first linear program that he constructed recommended daily consumption of several liters of vinegar. When he removed vinegar from the next input, he obtained approximately 200 bouillon cubes as the basis of the daily diet. This story, whose truth is not entirely out of the question, doesn't diminish the power of linear programming in any way, but it illustrates how difficult it is to capture mathematically all the important aspects of real-life problems. In the realm of nutrition, for example, it is not clear even today what exactly the influence of various components of food is on the human body. (Although, of course, many things *are* clear, and hopes that the science of the future will recommend hamburgers as the main ingredient of a healthy diet will almost surely be disappointed.) Even if it were known perfectly, few people want and can formulate exactly what they expect from their diet—apparently, it is much easier to formulate such requirements for the diet of someone else. Moreover, there are nonlinear dependencies among the effects of various nutrients, and so the dietary problem can never be captured perfectly by linear programming.

There are many applications of linear programming in industry, agriculture, services, etc. that from an abstract point of view are variations of the diet problem and do not introduce substantially new mathematical tricks. It may still be challenging to design good models for real-life problems of this kind, but the challenges are not mathematical. We will not dwell on such problems here (many examples can be found in Chvátal's book cited in Chapter 9), and we will present problems in which the use of linear programming has different flavors.

2.2 Flow in a Network

An administrator of a computer network convinced his employer to purchase a new computer with an improved sound system. He wants to transfer his music collection from an old computer to the new one, using a local network. The network looks like this:

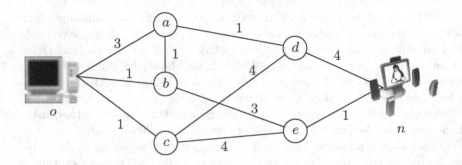

What is the maximum transfer rate from computer o (old) to computer n (new)? The numbers near each data link specify the maximum transfer rate of that link (in Mbit/s, say). We assume that each link can transfer data in either direction, but not in both directions simultaneously. So, for example, through the link ab one can *either* send data from a to b at any rate from 0 up to 1 Mbit/s, *or* send data from b to a at any rate from 0 to 1 Mbit/s.

The nodes $a, b, \ldots, e$ are not suitable for storing substantial amounts of data, and hence all data entering them has to be sent further immediately. From this we can already see that the maximum transfer rate cannot be used on all links simultaneously (consider node a, for example). Thus we have to find an appropriate value of the data flow for each link so that the total transfer rate from o to n is maximum.

For every link in the network we introduce one variable. For example, x_{be} specifies the rate by which data is transfered from b to e. Here x_{be} can also be negative, which means that data flow in the opposite direction, from e to b. (And we thus do *not* introduce another variable x_{eb}, which would correspond to the transfer rate from e to b.) There are 10 variables: x_{oa}, x_{ob}, x_{oc}, x_{ab}, x_{ad}, x_{be}, x_{cd}, x_{ce}, x_{dn}, and x_{en}.

We set up the following linear program:

Maximize $x_{oa} + x_{ob} + x_{oc}$

subject to $-3 \leq x_{oa} \leq 3, \quad -1 \leq x_{ob} \leq 1, \quad -1 \leq x_{oc} \leq 1$
$-1 \leq x_{ab} \leq 1, \quad -1 \leq x_{ad} \leq 1, \quad -3 \leq x_{be} \leq 3$
$-4 \leq x_{cd} \leq 4, \quad -4 \leq x_{ce} \leq 4, \quad -4 \leq x_{dn} \leq 4$
$-1 \leq x_{en} \leq 1$

$$x_{oa} = x_{ab} + x_{ad}$$
$$x_{ob} + x_{ab} = x_{be}$$
$$x_{oc} = x_{cd} + x_{ce}$$
$$x_{ad} + x_{cd} = x_{dn}$$
$$x_{be} + x_{ce} = x_{en}.$$

The objective function $x_{oa} + x_{ob} + x_{oc}$ expresses the total rate by which data is sent out from computer o. Since it is neither stored nor lost (hopefully) anywhere, it has to be received at n at the same rate. The next 10 constraints, $-3 \leq x_{oa} \leq 3$ through $-1 \leq x_{en} \leq 1$, restrict the transfer rates along the individual links. The remaining constraints say that whatever enters each of the nodes a through e has to leave immediately.

The optimal solution of this linear program is depicted below:

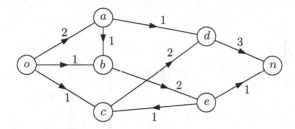

The number near each link is the transfer rate on that link, and the arrow determines the direction of the data flow. Note that between c and e data has to be sent in the direction from e to c, and hence $x_{ce} = -1$. The optimum value of the objective function is 4, and this is the desired maximum transfer rate.

In this example it is easy to see that the transfer rate cannot be larger, since the total capacity of all links connecting the computers o and a to the rest of the network equals 4. This is a special case of a remarkable theorem on maximum flow and minimum cut, which is usually discussed in courses on graph algorithms (see also Section 8.2).

Our example of data flow in a network is small and simple. In practice, however, flows are considered in intricate networks, sometimes even with many source nodes and sink nodes. These can be electrical networks (current flows), road or railroad networks (cars or trains flow), telephone networks (voice or data signals flow), financial (money flows), and so on. There are also many less-obvious applications of network flows—for example, in image processing.

Historically, the network flow problem was first formulated by American military experts in search of efficient ways of disrupting the railway system of the Soviet block; see

A. Schrijver: On the history of the transportation and maximum flow problems, *Math. Programming Ser. B* 91(2002) 437–445.

2.3 Ice Cream All Year Round

The next application of linear programming again concerns food (which should not be surprising, given the importance of food in life and the difficulties in optimizing sleep or love). The ice cream manufacturer Icicle Works Ltd.[2] needs to set up a production plan for the next year. Based on history, extensive surveys, and bird observations, the marketing department has come up with the following prediction of monthly sales of ice cream in the next year:

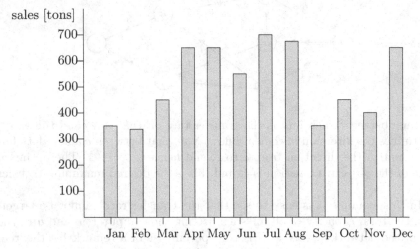

Now Icicle Works Ltd. needs to set up a production schedule to meet these demands.

A simple solution would be to produce "just in time," meaning that all the ice cream needed in month i is also produced in month i, $i = 1, 2, \ldots, 12$. However, this means that the produced amount would vary greatly from month to month, and a change in the produced amount has significant costs: Temporary workers have to be hired or laid off, machines have to be adjusted,

---

[2] Not to be confused with a rock group of the same name. The name comes from a nice science fiction story by Frederik Pohl.

and so on. So it would be better to spread the production more evenly over the year: In months with low demand, the idle capacities of the factory could be used to build up a stock of ice cream for the months with high demand.

So another simple solution might be a completely "flat" production schedule, with the same amount produced every month. Some thought reveals that such a schedule need not be feasible if we want to end up with zero surplus at the end of the year. But even if it is feasible, it need not be ideal either, since storing ice cream incurs a nontrivial cost. It seems likely that the best production schedule should be somewhere between these two extremes (production following demand and constant production). We want a compromise minimizing the total cost resulting both from changes in production and from storage of surpluses.

To formalize this problem, let us denote the demand in month i by $d_i \geq 0$ (in tons). Then we introduce a nonnegative variable x_i for the production in month i and another nonnegative variable s_i for the total surplus in store at the end of month i. To meet the demand in month i, we may use the production in month i and the surplus at the end of month $i - 1$:

$$x_i + s_{i-1} \geq d_i \quad \text{for } i = 1, 2, \ldots, 12.$$

The quantity $x_i + s_{i-1} - d_i$ is exactly the surplus after month i, and thus we have

$$x_i + s_{i-1} - s_i = d_i \quad \text{for } i = 1, 2, \ldots, 12.$$

Assuming that initially there is no surplus, we set $s_0 = 0$ (if we took the production history into account, s_0 would be the surplus at the end of the previous year). We also set $s_{12} = 0$, unless we want to plan for another year.

Among all nonnegative solutions to these equations, we are looking for one that minimizes the total cost. Let us assume that changing the production by 1 ton from month $i - 1$ to month i costs €50, and that storage facilities for 1 ton of ice cream cost €20 per month. Then the total cost is expressed by the function

$$50 \sum_{i=1}^{12} |x_i - x_{i-1}| + 20 \sum_{i=1}^{12} s_i,$$

where we set $x_0 = 0$ (again, history can easily be taken into account).

Unfortunately, this cost function is not linear. Fortunately, there is a simple but important trick that allows us to make it linear, at the price of introducing extra variables.

The change in production is either an increase or a decrease. Let us introduce a nonnegative variable y_i for the increase from month $i - 1$ to month i, and a nonnegative variable z_i for the decrease. Then

$$x_i - x_{i-1} = y_i - z_i \text{ and } |x_i - x_{i-1}| = y_i + z_i.$$

A production schedule of minimum total cost is given by an optimal solution of the following linear program:

$$\begin{aligned}
\text{Minimize} \quad & 50 \sum_{i=1}^{12} y_i + 50 \sum_{i=1}^{12} z_i + 20 \sum_{i=1}^{12} s_i \\
\text{subject to} \quad & x_i + s_{i-1} - s_i = d_i \text{ for } i = 1, 2, \ldots, 12 \\
& x_i - x_{i-1} = y_i - z_i \text{ for } i = 1, 2, \ldots, 12 \\
& x_0 = 0 \\
& s_0 = 0 \\
& s_{12} = 0 \\
& x_i, s_i, y_i, z_i \geq 0 \text{ for } i = 1, 2, \ldots, 12.
\end{aligned}$$

To see that an optimal solution $(\mathbf{s}^*, \mathbf{y}^*, \mathbf{z}^*)$ of this linear program indeed defines a schedule, we need to note that one of y_i^* and z_i^* has to be zero for all i, for otherwise, we could decrease both and obtain a better solution. This means that $y_i^* + z_i^*$ indeed equals the change in production from month $i - 1$ to month i, as required.

In the Icicle Works example above, this linear program yields the following production schedule (shown with black bars; the gray background graph represents the demands).

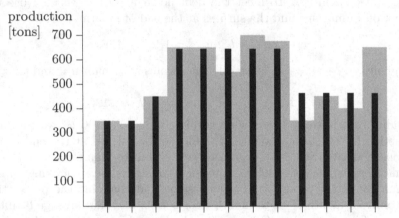

Below is the schedule we would get with zero storage costs (that is, after replacing the "20" by "0" in the above linear program).

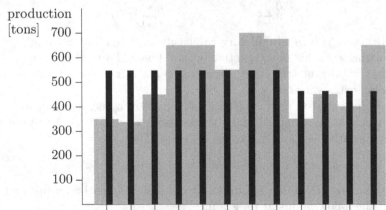

The pattern of this example is quite general, and many problems of optimal control can be solved via linear programming in a similar manner. A neat example is "Moon Rocket Landing," a once-popular game for programmable calculators (probably not sophisticated enough to survive in today's competition). A lunar module with limited fuel supply is descending vertically to the lunar surface under the influence of gravitation, and at chosen time intervals it can flash its rockets to slow down the descent (or even to start flying upward). The goal is to land on the surface with (almost) zero speed before exhausting all of the fuel. The reader is invited to formulate an appropriate linear program for determining the minimum amount of fuel necessary for landing, given the appropriate input data. For the linear programming formulation, we have to discretize time first (in the game this was done anyway), but with short enough time steps this doesn't make a difference in practice.

Let us remark that this particular problem can be solved analytically, with some calculus (or even mathematical control theory). But in even slightly more complicated situations, an analytic solution is out of reach.

2.4 Fitting a Line

The loudness level of nightingale singing was measured every evening for a number of days in a row, and the percentage of people watching the principal TV news was surveyed by questionnaires. The following diagram plots the measured values by points in the plane:

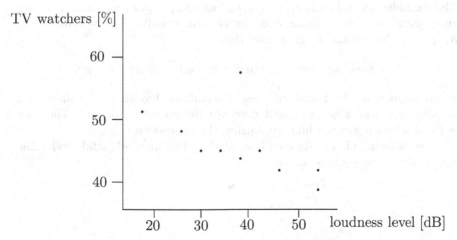

The simplest dependencies are linear, and many dependencies can be well approximated by a linear function. We thus want to find a line that best fits the measured points. (Readers feeling that this example is not sufficiently realistic can recall some measurements in physics labs, where the measured quantities should actually obey an exact linear dependence.)

How can one formulate mathematically that a given line "best fits" the points? There is no unique way, and several different criteria are commonly used for line fitting in practice.

The most popular one is the method of *least squares*, which for given points $(x_1, y_1), \ldots, (x_n, y_n)$ seeks a line with equation $y = ax + b$ minimizing the expression

$$\sum_{i=1}^{n}(ax_i + b - y_i)^2.\tag{2.1}$$

In words, for every point we take its vertical distance from the line, square it, and sum these "squares of errors."

This method need not always be the most suitable. For instance, if a few exceptional points are measured with very large error, they can influence the resulting line a great deal. An alternative method, less sensitive to a small number of "outliers," is to minimize the sum of absolute values of all errors:

$$\sum_{i=1}^{n}|ax_i + b - y_i|.\tag{2.2}$$

By a trick similar to the one we have seen in Section 2.3, this apparently nonlinear optimization problem can be captured by a linear program:

$$
\begin{aligned}
\text{Minimize} \quad & e_1 + e_2 + \cdots + e_n \\
\text{subject to} \quad & e_i \geq ax_i + b - y_i & \text{for } i = 1, 2, \ldots, n \\
& e_i \geq -(ax_i + b - y_i) & \text{for } i = 1, 2, \ldots, n.
\end{aligned}
$$

The variables are a, b, and $e_1, e_2, \ldots, e_n$ (while $x_1, \ldots, x_n$ and $y_1, \ldots, y_n$ are given numbers). Each e_i is an auxiliary variable standing for the error at the ith point. The constraints guarantee that

$$e_i \geq \max\Big(ax_i + b - y_i, -(ax_i + b - y_i)\Big) = |ax_i + b - y_i|.$$

In an optimal solution each of these inequalities has to be satisfied with equality, for otherwise, we could decrease the corresponding e_i. Thus, an optimal solution yields a line minimizing the expression (2.2).

The following picture shows a line fitted by this method (solid) and a line fitted using least squares (dotted):

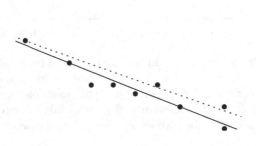

In conclusion, let us recall the useful trick we have learned here and in the previous section:

> Objective functions or constraints involving absolute values can often be handled via linear programming by introducing extra variables or extra constraints.

2.5 Separation of Points

A computer-controlled rabbit trap "Gromit RT 2.1" should be programmed so that it catches rabbits, but if a weasel wanders in, it is released. The trap can weigh the animal inside and also can determine the area of its shadow.

These two parameters were collected for a number of specimens of rabbits and weasels, as depicted in the following graph:

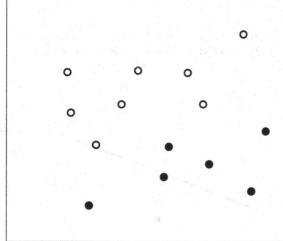

shadow area

(empty circles represent rabbits and full circles weasels).

Apparently, neither weight alone nor shadow area alone can be used to tell a rabbit from a weasel. One of the next-simplest things would be a linear criterion distinguishing them. That is, geometrically, we would like to separate the black points from the white points by a straight line if possible. Mathematically speaking, we are given m white points $\mathbf{p}_1, \mathbf{p}_2, \ldots, \mathbf{p}_m$

and n black points $\mathbf{q}_1, \mathbf{q}_2, \ldots, \mathbf{q}_n$ in the plane, and we would like to find out whether there exists a line having all white points on one side and all black points on the other side (none of the points should lie on the line).

In a solution of this problem by linear programming we distinguish three cases. First we test whether there exists a *vertical* line with the required property. This case needs neither linear programming nor particular cleverness.

The next case is the existence of a line that is not vertical and that has all black points below it and all white points above it. Let us write the equation of such a line as $y = ax + b$, where a and b are some yet unknown real numbers. A point $\mathbf{r}$ with coordinates $x(\mathbf{r})$ and $y(\mathbf{r})$ lies above this line if $y(\mathbf{r}) > ax(\mathbf{r}) + b$, and it lies below it if $y(\mathbf{r}) < ax(\mathbf{r}) + b$. So a suitable line exists if and only if the following system of inequalities with variables a and b has a solution:

$$y(\mathbf{p}_i) > ax(\mathbf{p}_i) + b \qquad \text{for } i = 1, 2, \ldots, m$$
$$y(\mathbf{q}_j) < ax(\mathbf{q}_j) + b \qquad \text{for } j = 1, 2, \ldots, n.$$

We haven't yet mentioned strict inequalities in connection with linear programming, and actually, they are not allowed in linear programs. But here we can get around this issue by a small trick: We introduce a new variable δ, which stands for the "gap" between the left and right sides of each strict inequality. Then we try to make the gap as large as possible:

$$\text{Maximize} \quad \delta$$
$$\text{subject to} \quad y(\mathbf{p}_i) \geq ax(\mathbf{p}_i) + b + \delta \quad \text{for } i = 1, 2, \ldots, m$$
$$y(\mathbf{q}_j) \leq ax(\mathbf{q}_j) + b - \delta \quad \text{for } j = 1, 2, \ldots, n.$$

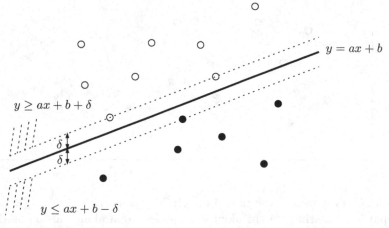

This linear program has three variables: a, b, and δ. The optimal δ is positive exactly if the preceding system of strict inequalities has a solution, and the latter happens exactly if a nonvertical line exists with all black points below and all white points above.

Similarly, we can deal with the third case, namely the existence of a non-vertical line having all black points above it and all white points below it. This completes the description of an algorithm for the line separation problem.

A plane separating two point sets in $\mathbb{R}^3$ can be computed by the same approach, and we can also solve the analogous problem in higher dimensions. So we could try to distinguish rabbits from weasels based on more than two measured parameters.

Here is another, perhaps more surprising, extension. Let us imagine that separating rabbits from weasels by a straight line proved impossible. Then we could try, for instance, separating them by a graph of a quadratic function (a parabola), of the form $ax^2 + bx + c$. So given m white points $\mathbf{p}_1, \mathbf{p}_2, \ldots, \mathbf{p}_m$ and n black points $\mathbf{q}_1, \mathbf{q}_2, \ldots, \mathbf{q}_n$ in the plane, we now ask, are there coefficients $a, b, c \in \mathbb{R}$ such that the graph of $f(x) = ax^2 + bx + c$ has all white points above it and all black points below? This leads to the inequality system

$$y(\mathbf{p}_i) > ax(\mathbf{p}_i)^2 + bx(\mathbf{p}_i) + c \qquad \text{for } i = 1, 2, \ldots, m$$
$$y(\mathbf{q}_j) < ax(\mathbf{q}_j)^2 + bx(\mathbf{q}_j) + c \qquad \text{for } j = 1, 2, \ldots, n.$$

By introducing a gap variable δ as before, this can be written as the following linear program in the variables a, b, c, and δ:

Maximize $\quad \delta$
subject to $\quad y(\mathbf{p}_i) \geq ax(\mathbf{p}_i)^2 + bx(\mathbf{p}_i) + c + \delta \qquad \text{for } i = 1, 2, \ldots, m$
$\qquad\qquad\quad y(\mathbf{q}_j) \leq ax(\mathbf{q}_j)^2 + bx(\mathbf{q}_j) + c - \delta \qquad \text{for } j = 1, 2, \ldots, n.$

In this linear program the quadratic terms are coefficients and therefore they cause no harm.

The same approach also allows us to test whether two point sets in the plane, or in higher dimensions, can be separated by a function of the form $f(\mathbf{x}) = a_1\varphi_1(\mathbf{x}) + a_2\varphi_2(\mathbf{x}) + \cdots + a_k\varphi_k(\mathbf{x})$, where $\varphi_1, \ldots, \varphi_k$ are given functions (possibly nonlinear) and $a_1, a_2, \ldots, a_k$ are real coefficients, in the sense that $f(\mathbf{p}_i) > 0$ for every white point $\mathbf{p}_i$ and $f(\mathbf{q}_j) < 0$ for every black point $\mathbf{q}_j$.

2.6 Largest Disk in a Convex Polygon

Here we will encounter another problem that may look nonlinear at first sight but can be transformed to a linear program. It is a simple instance of a geometric *packing problem*: Given a container, in our case a convex polygon, we want to fit as large an object as possible into it, in our case a disk of the largest possible radius.

Let us call the given convex polygon P, and let us assume that it has n sides. As we said, we want to find the largest circular disk contained in P.

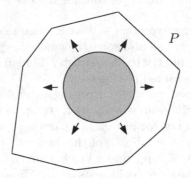

For simplicity let us assume that none of the sides of P is vertical. Let the ith side of P lie on a line ℓ_i with equation $y = a_i x + b_i$, $i = 1, 2, \ldots, n$, and let us choose the numbering of the sides in such a way that the first, second, up to the kth side bound P from below, while the $(k+1)$st through nth side bound it from above.

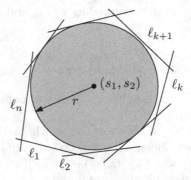

Let us now ask, under what conditions does a circle with center $\mathbf{s} = (s_1, s_2)$ and radius r lie completely inside P? This is the case if and only if the point $\mathbf{s}$ has distance at least r from each of the lines $\ell_1, \ldots, \ell_n$, lies above the lines $\ell_1, \ldots, \ell_k$, and lies below the lines $\ell_{k+1}, \ldots, \ell_n$. We compute the distance of $\mathbf{s}$ from ℓ_i. A simple calculation using similarity of triangles and the Pythagorean theorem shows that this distance equals the absolute value of the expression

$$\frac{s_2 - a_i s_1 - b_i}{\sqrt{a_i^2 + 1}}.$$

Moreover, the expression is positive if $\mathbf{s}$ lies above ℓ_i, and it is negative if $\mathbf{s}$ lies below ℓ_i:

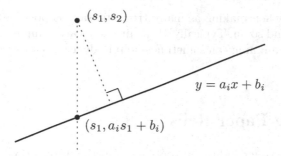

The disk of radius r centered at $\mathbf{s}$ thus lies inside P exactly if the following system of inequalities is satisfied:

$$\frac{s_2 - a_i s_1 - b_i}{\sqrt{a_i^2 + 1}} \geq r, \qquad i = 1, 2, \ldots, k$$

$$\frac{s_2 - a_i s_1 - b_i}{\sqrt{a_i^2 + 1}} \leq -r, \qquad i = k+1, k+2, \ldots, n.$$

Therefore, we want to find the largest r such that there exist s_1 and s_2 so that all the constraints are satisfied. This yields a linear program! (Some might be frightened by the square roots, but these can be computed in advance, since all the a_i are concrete numbers.)

$$\text{Maximize} \quad r$$

$$\text{subject to} \quad \frac{s_2 - a_i s_1 - b_i}{\sqrt{a_i^2 + 1}} \geq r \quad \text{for } i = 1, 2, \ldots, k$$

$$\frac{s_2 - a_i s_1 - b_i}{\sqrt{a_i^2 + 1}} \leq -r \quad \text{for } i = k+1, k+2, \ldots, n.$$

There are three variables: s_1, s_2, and r. An optimal solution yields the desired largest disk contained in P.

A similar problem in higher dimension can be solved analogously. For example, in three-dimensional space we can ask for the largest ball that can be placed into the intersection of n given half-spaces.

Interestingly, another similar-looking problem, namely, finding the smallest disk containing a given convex n-gon in the plane, cannot be expressed by a linear program and has to be solved differently; see Section 8.7.

Both in practice and in theory, one usually encounters geometric packing problems that are more complicated than the one considered in this section and not so easily solved by linear programming. Often we have a fixed collection of objects and we want to pack as many of them as possible into a given container (or several containers). Such problems are encountered by confectioners when cutting cookies from a piece of dough, by tailors or clothing manufacturers when making as many trousers, say, as possible from a large piece of cloth, and so on. Typically, these problems are computationally hard, but linear programming can sometimes help in devising heuristics or approximate algorithms.

2.7 Cutting Paper Rolls

Here we have another industrial problem, and the application of linear programming is quite nonobvious. Moreover, we will naturally encounter an integrality constraint, which will bring us to the topic of the next chapter.

A paper mill manufactures rolls of paper of a standard width 3 meters. But customers want to buy paper rolls of shorter width, and the mill has to cut such rolls from the 3 m rolls. One 3 m roll can be cut, for instance, into two rolls 93 cm wide, one roll of width 108 cm, and a rest of 6 cm (which goes to waste).

Let us consider an order of

- 97 rolls of width 135 cm,
- 610 rolls of width 108 cm,
- 395 rolls of width 93 cm, and
- 211 rolls of width 42 cm.

What is the smallest number of 3 m rolls that have to be cut in order to satisfy this order, and how should they be cut?

In order to engage linear programming one has to be generous in introducing variables. We write down all of the requested widths: 135 cm, 108 cm, 93 cm, and 42 cm. Then we list all possibilities of cutting a 3 m paper roll into rolls of some of these widths (we need to consider only possibilities for which the wasted piece is shorter than 42 cm):

P1:	2×135	P7:	$108 + 93 + 2 \times 42$
P2:	$135 + 108 + 42$	P8:	$108 + 4 \times 42$
P3:	$135 + 93 + 42$	P9:	3×93
P4:	$135 + 3 \times 42$	P10:	$2 \times 93 + 2 \times 42$
P5:	$2 \times 108 + 2 \times 42$	P11:	$93 + 4 \times 42$
P6:	$108 + 2 \times 93$	P12:	7×42

For each possibility Pj on the list we introduce a variable $x_j \geq 0$ representing the number of rolls cut according to that possibility. We want to minimize the total number of rolls cut, i.e., $\sum_{j=1}^{12} x_j$, in such a way that the customers are satisfied. For example, to satisfy the demand for 395 rolls of width 93 cm we require

$$x_3 + 2x_6 + x_7 + 3x_9 + 2x_{10} + x_{11} \geq 395.$$

For each of the widths we obtain one constraint.

For a more complicated order, the list of possibilities would most likely be produced by computer. We would be in a quite typical situation in which a linear program is not entered "by hand," but rather is generated by some computer program. More-advanced techniques even generate the possibilities "on the fly," during the solution of the linear program, which may save time and memory considerably. See the entry "column generation" in the glossary or Chvátal's book cited in Chapter 9, from which this example is taken.

The optimal solution of the resulting linear program has $x_1 = 48.5$, $x_5 = 206.25$, $x_6 = 197.5$, and all other components 0. In order to cut 48.5 rolls according to the possibility P1, one has to unwind half of a roll. Here we need more information about the technical possibilities of the paper mill: Is cutting a fraction of a roll technically and economically feasible? If yes, we have solved the problem optimally. If not, we have to work further and somehow take into account the restriction that only feasible solutions of the linear program with *integral* x_i are of interest. This is not at all easy in general, and it is the subject of Chapter 3.

3. Integer Programming and LP Relaxation

3.1 Integer Programming

In Section 2.7 we encountered a situation in which among all feasible solutions of a linear program, only those with all components integral are of interest in the practical application. A similar situation occurs quite often in attempts to apply linear programming, because objects that can be split into arbitrary fractions are more an exception than the rule. When hiring workers, scheduling buses, or cutting paper rolls one somehow has to deal with the fact that workers, buses, and paper rolls occur only in integral quantities.

Sometimes an optimal or almost-optimal integral solution can be obtained by simply rounding the components of an optimal solution of the linear program to integers, either up, or down, or to the nearest integer. In our paper-cutting example from Section 2.7 it is natural to round up, since we have to fulfill the order. Starting from the optimal solution $x_1 = 48.5$, $x_5 = 206.25$, $x_6 = 197.5$ of the linear program, we thus arrive at the integral solution $x_1 = 49$, $x_5 = 207$, and $x_6 = 198$, which means cutting 454 rolls. Since we have found an optimum of the linear program, we know that no solution whatsoever, even one with fractional amounts of rolls allowed, can do better than cutting 452.5 rolls. If we insist on cutting an integral number of rolls, we can thus be sure that at least 453 rolls must be cut. So the solution obtained by rounding is quite good.

However, it turns out that we can do slightly better. The integral solution $x_1 = 49$, $x_5 = 207$, $x_6 = 196$, and $x_9 = 1$ (with all other components 0) requires cutting only 453 rolls. By the above considerations, no integral solution can do better.

In general, the gap between a rounded solution and an optimal integral solution can be much larger. If the linear program specifies that for most of 197 bus lines connecting villages it is best to schedule something between 0.1 and 0.3 buses, then, clearly, rounding to integers exerts a truly radical influence.

The problem of cutting paper rolls actually leads to a problem with a linear objective function and linear constraints (equations and inequalities), but the variables are allowed to attain only integer values. Such an optimization problem is called an integer program, and after a small adjustment we can write it in a way similar to that used for a linear program in Chapter 1:

An integer program:

Maximize $\mathbf{c}^T\mathbf{x}$
subject to $A\mathbf{x} \leq \mathbf{b}$
            $\mathbf{x} \in \mathbb{Z}^n$.

Here A is an $m \times n$ matrix, $\mathbf{b} \in \mathbb{R}^m$, $\mathbf{c} \in \mathbb{R}^n$, $\mathbb{Z}$ denotes the set of integers, and $\mathbb{Z}^n$ is the set of all n-component integer vectors.

The set of all feasible solutions of an integer program is no longer a convex polyhedron, as was the case for linear programming, but it consists of separate integer points. A picture illustrates a two-dimensional integer program with five constraints:

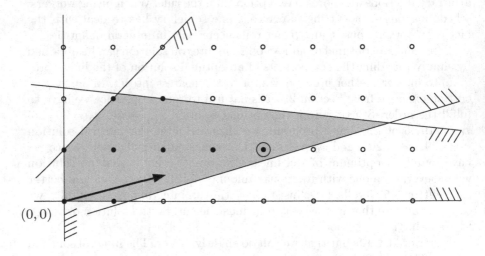

Feasible solutions are shown as solid dots and the optimal solution is marked by a circle. Note that it lies quite far from the optimum of the *linear* program with the same five constraints and the same objective function.

It is known that *solving a general integer program is computationally difficult* (more exactly, it is an NP-*hard problem*), in contrast to solving a linear program. Linear programs with many thousands of variables and constraints can be handled in practice, but there are integer programs with 10 variables and 10 constraints that are insurmountable even for the most modern computers and software.

Adding the integrality constraints can thus change the difficulty of a problem in a drastic way indeed. This may not look so surprising anymore if we realize that integer programs can model yes/no decisions, since an integer variable x_j satisfying the linear constraints $0 \leq x_j \leq 1$ has possible values only 0 (no) and 1 (yes). For those familiar with the foundations of NP-completeness it is thus not hard to model the problem of satisfiability of logical formulas by an integer program. In Section 3.4 we will see how an integer program can express the maximum size of an independent set in a given graph, which is also one of the basic NP-hard problems.

Several techniques have been developed for solving integer programs. In the literature, some of them can be found under the headings *cutting planes*, *branch and bound*, as well as *branch and cut* (see the glossary). The most successful strategies usually employ linear programming as a subroutine for solving certain auxiliary problems. How to do this efficiently is investigated in a branch of mathematics called *polyhedral combinatorics*.

The most widespread use of linear programming today, and the one that consumes the largest share of computer time, is most likely in auxiliary computations for integer programs.

Let us remark that there are many optimization problems in which some of the variables are integral, while others may attain arbitrary real values. Then one speaks of *mixed integer programming*. This is in all likelihood the most frequent type of optimization problem in practice.

We will demonstrate several important optimization problems that can easily be formulated as integer programs, and we will show how linear programming can or cannot be used in their solution. But it will be only a small sample from this area, which has recently developed extensively and which uses many complicated techniques and clever tricks.

3.2 Maximum-Weight Matching

A consulting company underwent a thorough reorganization, in order to adapt to current trends, in which the department of Creative Accounting with 7 employees was closed down. But flexibly enough, seven new positions have been created. The human resources manager, in order to assign the new positions to the seven employees, conducted interviews with them and gave them extensive questionnaires to fill out. Then he summarized the results in scores: Each employee got a score between 0 and 100 for each of the positions she or he was willing to accept. The manager depicted this information in a diagram, in which an expert can immediately recognize a bipartite graph:

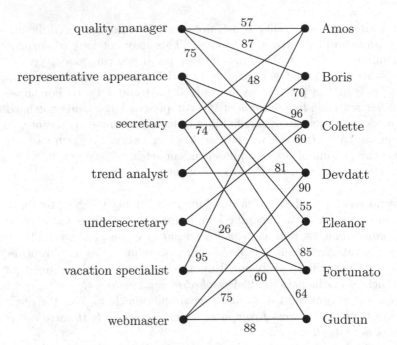

For example, this diagram tells us that Boris is willing to accept the job in quality management, for which he achieved score of 87, or the job of a trend analyst, for which he has score 70. Now the manager wants to select a position for everyone so that the sum of scores is maximized. The first idea naturally coming to mind is to give everyone the position for which he/she has the largest score. But this cannot be done, since, for example, three people are best suited for the profession of webmaster: Eleanor, Gudrun, and Devdatt. If we try to assign the positions by a "greedy" algorithm, meaning that in each step we make an assignment of largest possible score between a yet unoccupied position and a still unassigned employee, we end up with filling only 6 positions:

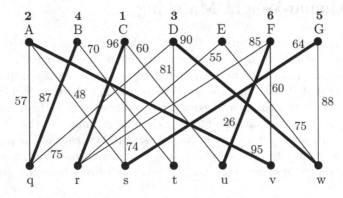

(The bold digits 1–6 indicate the order of making assignments by the greedy algorithm.)

In the language of graph theory we have a bipartite graph with vertex set $V = X \cup Y$ and edge set E. Each edge connects a vertex of X to a vertex of Y. Moreover, we have $|X| = |Y|$. For each edge $e \in E$ we are given a nonnegative weight w_e. We want to find a subset $M \subseteq E$ of edges such that each vertex of both X and Y is incident to exactly one edge of M (such an M is called a **perfect matching**), and the sum $\sum_{e \in M} w_e$ is the largest possible.

In order to formulate this problem as an integer program, we introduce variables x_e, one for each edge $e \in E$, that can attain values 0 or 1. They will encode the sought-after set M: $x_e = 1$ means $e \in M$ and $x_e = 0$ means $e \notin M$. Then $\sum_{e \in M} w_e$ can be written as

$$\sum_{e \in E} w_e x_e,$$

and this is the objective function. The requirement that a vertex $v \in V$ have exactly one incident edge of M is expressed by having the sum of x_e over all edges incident to v equal to 1. In symbols, $\sum_{e \in E: v \in e} x_e = 1$. The resulting integer program is

$$\begin{aligned}
&\text{maximize} && \textstyle\sum_{e \in E} w_e x_e \\
&\text{subject to} && \textstyle\sum_{e \in E: v \in e} x_e = 1 \text{ for each vertex } v \in V, \text{ and} \\
& && x_e \in \{0, 1\} \text{ for each edge } e \in E.
\end{aligned} \qquad (3.1)$$

If we leave out the integrality conditions, i.e., if we allow each x_e to attain all values in the interval $[0, 1]$, we obtain the following linear program:

$$\begin{aligned}
&\text{Maximize} && \textstyle\sum_{e \in E} w_e x_e \\
&\text{subject to} && \textstyle\sum_{e \in E: v \in e} x_e = 1 \text{ for each vertex } v \in V, \text{ and} \\
& && 0 \leq x_e \leq 1 \text{ for each edge } e \in E.
\end{aligned}$$

It is called an **LP relaxation** of the integer program (3.1)—we have relaxed the constraints $x_e \in \{0, 1\}$ to the weaker constraints $0 \leq x_e \leq 1$. We can solve the LP relaxation, say by the simplex method, and either we obtain an optimal solution $\mathbf{x}^*$, or we learn that the LP relaxation is infeasible. In the latter case, the original integer program must be infeasible as well, and consequently, there is no perfect matching.

Let us now assume that the LP relaxation has an optimal solution $\mathbf{x}^*$. What can such an $\mathbf{x}^*$ be good for? Certainly it provides an *upper bound* on the best possible solution of the original integer program (3.1). More precisely, the optimum of the objective function in the integer program (3.1) is bounded above by the value of the objective function at $\mathbf{x}^*$. This is because every feasible solution of the integer program is also a feasible solution of the LP relaxation, and so we are maximizing over a larger set of vectors in the LP relaxation. An upper bound can be very valuable: For example, if we manage to find a feasible solution of the integer program for which the value of the objective function is 98% of the upper bound, we can usually stop striving, since we know that we cannot improve by more than roughly 2% no matter how hard we try. (Of course, it depends on what we are dealing with; if it is a state budget, even 2% is still worth some effort.)

A pleasant surprise awaits us in the particular problem we are considering here: The LP relaxation not only yields an upper bound, but it provides an optimal solution of the original problem! Namely, if we solve the LP relaxation by the simplex method, we obtain an optimal $\mathbf{x}^*$ that has all components equal to 0 or 1, and thus it determines an optimum perfect matching. (If a better perfect matching existed, it would determine a better solution of the LP relaxation.) The optimal solution discovered in this way is drawn below:

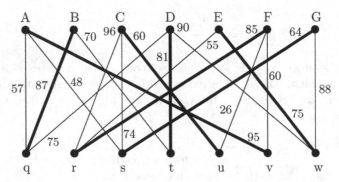

The following (nontrivial) theorem shows that things work this nicely for every problem of the considered type.

3.2.1 Theorem. *Let $G = (V, E)$ be an arbitrary bipartite graph with real edge weights w_e. If the LP relaxation of the integer program (3.1) has at least one feasible solution, then it has at least one integral optimal solution. This is an optimal solution for the integer program (3.1) as well.*

An interested reader can find a proof at the end of this section.

The theorem doesn't say that *every* optimal solution of the LP relaxation is necessarily integral. However, the proof gives an easy recipe for producing an integral optimal solution from an arbitrary optimal solution. Moreover, it can be shown (see Section 8.2) that the simplex method always returns an integral optimal solution (using the terminology of Chapter 4, each basic feasible solution is integral).

An LP relaxation can be considered for an arbitrary integer program: The condition $\mathbf{x} \in \mathbb{Z}^n$ is simply replaced by $\mathbf{x} \in \mathbb{R}^n$. Cases in which we always get an optimal solution of the integer program from the LP relaxation, such as in Theorem 3.2.1, are rather rare, but LP relaxation can also be useful in other ways.

The maximum of the objective function for the LP relaxation always provides an upper bound for the maximum of the integer program. Sometimes this bound is quite tight, but at other times it can be very bad; see Section 3.4. The quality of the upper bound from an LP relaxation has been studied for many types of problems. Sometimes one adds new linear constraints to the LP relaxation that are satisfied by all integral solutions (that is, all feasible solutions of the integer program satisfy these new constraints) but that exclude some of the nonintegral solutions. In this way the upper bound on the optimum of the integer program can often be greatly improved. If we continue adding suitable new constraints for long enough, then we even arrive at an optimal solution of the integer program. This is the main idea of the method of cutting planes.

A nonintegral optimal solution of the LP relaxation can sometimes be converted to an approximately optimal solution of the integer program by appropriate rounding. We will see a simple example in Section 3.3, and a more advanced one in Section 8.3.

Proof of Theorem 3.2.1. Let $\mathbf{x}^*$ be an optimal solution of the LP relaxation, and let $w(\mathbf{x}^*) = \sum_{e \in E} w_e x_e^*$ be the value of the objective function at $\mathbf{x}^*$. Let us denote the number of nonintegral components of the vector $\mathbf{x}^*$ by $k(\mathbf{x}^*)$.

If $k(\mathbf{x}^*) = 0$, then we are done. For $k(\mathbf{x}^*) > 0$ we describe a procedure that yields another optimal solution $\tilde{\mathbf{x}}$ with $k(\tilde{\mathbf{x}}) < k(\mathbf{x}^*)$. We reach an integral optimal solution by finitely many repetitions of this procedure.

Let $x_{e_1}^*$ be a nonintegral component of the vector $\mathbf{x}^*$, corresponding to some edge $e_1 = \{a_1, b_1\}$. Since $0 < x_{e_1}^* < 1$ and

$$\sum_{e \in E : b_1 \in e} x_e^* = 1,$$

there exists another edge $e_2 = \{a_2, b_1\}$, $a_2 \neq a_1$, with $x_{e_2}^*$ nonintegral. For a similar reason we can also find a third edge $e_3 = \{a_2, b_2\}$ with $0 < x_{e_3}^* < 1$. We continue in this manner and look for nonintegral components along a longer and longer path $(a_1, b_1, a_2, b_2, \ldots)$:

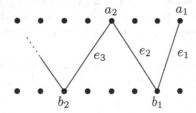

Since the graph G has finitely many vertices, eventually we reach a vertex that we have already visited before. This means that we have found a *cycle* $C \subseteq E$ in which $0 < x_e^* < 1$ for all edges. Since the graph is bipartite, the length t of the cycle C is even. For simplicity of notation let us assume that the edges of C are $e_1, e_2, \ldots, e_t$, although in reality the cycle found as above need not begin with the edge e_1.

Now for a small real number ε we define a vector $\tilde{\mathbf{x}}$ by

$$\tilde{x}_e = \begin{cases} x_e^* - \varepsilon & \text{for } e \in \{e_1, e_3, \ldots, e_{t-1}\} \\ x_e^* + \varepsilon & \text{for } e \in \{e_2, e_4, \ldots, e_t\} \\ x_e^* & \text{otherwise.} \end{cases}$$

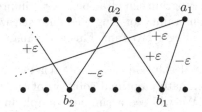

It is easy to see that this $\tilde{\mathbf{x}}$ satisfies all the conditions

$$\sum_{e \in E : v \in e} \tilde{x}_e = 1, \quad v \in V,$$

since at the vertices of the cycle C we have added ε once and subtracted it once, while for all other vertices the variables of the incident edges haven't changed their values at all. For ε sufficiently small the conditions $0 \leq \tilde{x}_e \leq 1$ are satisfied too, since all components $x_{e_i}^*$ are strictly between 0 and 1. Hence $\tilde{\mathbf{x}}$ is again a feasible solution of the LP relaxation for all sufficiently small ε (positive or negative).

What happens with the value of the objective function? We have

$$w(\tilde{\mathbf{x}}) = \sum_{e \in E} w_e \tilde{x}_e = w(\mathbf{x}^*) + \varepsilon \sum_{i=1}^{t} (-1)^i w_{e_i} = w(\mathbf{x}^*) + \varepsilon \Delta,$$

where we have set $\Delta = \sum_{i=1}^{t} (-1)^i w_{e_i}$. Since $\mathbf{x}^*$ is optimal, necessarily $\Delta = 0$, for otherwise, we could achieve $w(\tilde{\mathbf{x}}) > w(\mathbf{x}^*)$ either by choosing $\varepsilon > 0$ (for $\Delta > 0$) or by choosing $\varepsilon < 0$ (for $\Delta < 0$). This means that $\tilde{\mathbf{x}}$ is an optimal solution whenever it is feasible, i.e., for all ε with a sufficiently small absolute value.

Let us now choose the largest $\varepsilon > 0$ such that $\tilde{\mathbf{x}}$ is still feasible. Then there has to exist $e \in \{e_1, e_2, \ldots, e_t\}$ with $\tilde{x}_e \in \{0, 1\}$, and $\tilde{\mathbf{x}}$ has fewer nonintegral components than $\mathbf{x}^*$. □

Let us now consider another situation, in which we have more employees than positions and we want to fill all positions optimally, i.e., so that the sum of scores is maximized. Then we have $|X| < |Y|$

in the considered bipartite graph, and for vertices $v \in Y$ the condition $\sum_{e \in E: v \in e} x_e = 1$ (every vertex of Y is incident to *exactly one* edge of M) is replaced by $\sum_{e \in E: v \in e} x_e \leq 1$ (every vertex of Y is incident to *at most one* edge of M). The claim of Theorem 3.2.1 remains valid: If the LP relaxation has a feasible solution, then it also has an integral optimal solution. The proof presented above can be modified to show this. We present a different and more conceptual proof in Section 8.2 (which also yields an alternative proof of Theorem 3.2.1). We will also briefly discuss the nonbipartite case in Section 8.2.

3.3 Minimum Vertex Cover

The Internet had been expanding rapidly in the Free Republic of West Mordor, and the government issued a regulation, purely in the interest of improved security of the citizens, that every data link connecting two computers must be equipped with a special device for gathering statistical data about the traffic. An operator of a part of the network has to attach the government's monitoring boxes to some of his computers so that each link has a monitored computer on at least one end. Which computers should get boxes so that the total price is minimum? Let us assume that there is a flat rate per box.

It is again convenient to use graph-theoretic terminology. The computers in the network are vertices and the links are edges. So we have a graph $G = (V, E)$ and we want to find a subset $S \subseteq V$ of the vertices such that each edge has at least one end-vertex in S (such an S is called a **vertex cover**), and S is as small as possible.

This problem can be written as an integer program:

$$
\begin{aligned}
\text{Minimize} \quad & \sum_{v \in V} x_v \\
\text{subject to} \quad & x_u + x_v \geq 1 \quad \text{for every edge } \{u, v\} \in E \\
& x_v \in \{0, 1\} \quad \text{for all } v \in V.
\end{aligned}
\tag{3.2}
$$

For every vertex v we have a variable x_v, which can attain values 0 or 1. The meaning of $x_v = 1$ is $v \in S$, and $x_v = 0$ means $v \notin S$. The constraint $x_u + x_v \geq 1$ guarantees that the edge $\{u, v\}$ has at least one vertex in S. The objective function is the size of S.

It is known that finding a minimum vertex cover is a computationally difficult (NP-hard) problem. We will describe an approximation algorithm based on linear programming that always finds a vertex cover with at most twice as many vertices as in the smallest possible vertex cover.

An LP relaxation of the above integer program is

$$
\begin{aligned}
\text{minimize} \quad & \sum_{v \in V} x_v \\
\text{subject to} \quad & x_u + x_v \geq 1 \quad \text{for every edge } \{u, v\} \in E \\
& 0 \leq x_v \leq 1 \quad \text{for all } v \in V.
\end{aligned}
\tag{3.3}
$$

The first step of the approximation algorithm for vertex cover consists in computing an optimal solution $\mathbf{x}^*$ of this LP relaxation (by some standard algorithm for linear programming). The components of $\mathbf{x}^*$ are real numbers in the interval $[0, 1]$. In the second step we define the set

$$S_{\mathrm{LP}} = \{v \in V : x_v^* \geq \tfrac{1}{2}\}.$$

This is a vertex cover, since for every edge $\{u, v\}$ we have $x_u^* + x_v^* \geq 1$, and so $x_u^* \geq \tfrac{1}{2}$ or $x_v^* \geq \tfrac{1}{2}$.

Let S_{OPT} be some vertex cover of the minimum possible size (we don't have it but we can theorize about it). We claim that

$$|S_{\mathrm{LP}}| \leq 2 \cdot |S_{\mathrm{OPT}}|.$$

To see this, let $\tilde{\mathbf{x}}$ be an optimal solution of the integer program (3.2), which corresponds to the set S_{OPT}, i.e., $\tilde{x}_v = 1$ for $v \in S_{\mathrm{OPT}}$ and $\tilde{x}_v = 0$ otherwise. This $\tilde{\mathbf{x}}$ is definitely a feasible solution of the LP relaxation (3.3), and so it cannot have a smaller value of the objective function than an optimal solution $\mathbf{x}^*$ of (3.3):

$$\sum_{v \in V} x_v^* \leq \sum_{v \in V} \tilde{x}_v.$$

On the other hand, $|S_{\mathrm{LP}}| = \sum_{v \in S_{\mathrm{LP}}} 1 \leq \sum_{v \in V} 2x_v^*$, since $x_v^* \geq \tfrac{1}{2}$ for each $v \in S_{\mathrm{LP}}$. Therefore

$$|S_{\mathrm{LP}}| \leq 2 \cdot \sum_{v \in V} x_v^* \leq 2 \cdot \sum_{v \in V} \tilde{x}_v = 2 \cdot |S_{\mathrm{OPT}}|.$$

This proof illustrates an important aspect of approximation algorithms: In order to assess the quality of the computed solution, we always need a bound on the quality of the optimal solution, although we don't know it. The LP relaxation provides such a bound, which can sometimes be useful, as in the example of this section. In other problems it may be useless, though, as we will see in the next section.

Remarks. A natural attempt at an approximate solution of the considered problem is again a *greedy algorithm*: Select vertices one by one and always take a vertex that covers the maximum possible number of yet uncovered edges. Although this algorithm may not be bad in most cases, examples can be constructed in which it yields a solution at least ten times worse, say, than an optimal solution (and 10 can be replaced by any other constant). Discovering such a construction is a lovely exercise.

There is another, combinatorial, approximation algorithm for the minimum vertex cover: First we find a maximal matching M, that is, a matching that cannot be extended by adding any other edge (we note that such a matching need not have the maximum possible *number* of edges). Then we use the vertices covered by M as a vertex cover. This always gives a vertex cover at most twice as big as the optimum, similar to the algorithm explained above.

The algorithm based on linear programming has the advantage of being easy to generalize for a **weighted vertex cover** (the government boxes may have different prices for different computers). In the same way as we did for unit prices one can show that the cost of the computed solution is never larger than twice the optimum cost. As in the unweighted case, this result can also be achieved with combinatorial algorithms, but these are more difficult to understand than the linear programming approach.

3.4 Maximum Independent Set

Here the authors got tired of inventing imitations of real-life problems, and so we formulate the next problem in the language of graph theory right away. For a graph $G = (V, E)$, a set $I \subseteq V$ of vertices is called **independent** (or *stable*) if no two vertices of I are connected by an edge in G.

Computing an independent set with the maximum possible number of vertices for a given graph is one of the notoriously difficult graph-theoretic problems. It can be easily expressed by an integer program:

$$
\begin{aligned}
\text{Maximize} \quad & \sum_{v \in V} x_v \\
\text{subject to} \quad & x_u + x_v \leq 1 \quad \text{for each edge } \{u, v\} \in E, \text{ and} \\
& x_v \in \{0, 1\} \quad \text{for all } v \in V.
\end{aligned} \tag{3.4}
$$

An optimal solution $\mathbf{x}^*$ corresponds to a maximum independent set: $v \in I$ if and only if $x_v^* = 1$. The constraints $x_u + x_v \leq 1$ ensure that whenever two vertices u and v are connected by an edge, only one of them can get into I.

In an LP relaxation the condition $x_v \in \{0, 1\}$ is replaced by the inequalities $0 \leq x_v \leq 1$. The resulting linear program always has a feasible solution with all $x_v = \frac{1}{2}$, which yields objective function equal to $\frac{|V|}{2}$. Hence the optimal value of the objective function is $\frac{|V|}{2}$ or larger.

Let us consider a complete graph on n vertices (the graph in which every two vertices are connected by an edge). The largest independent set consists of a single vertex and thus has size 1. However, as we have seen, the optimal value for the LP relaxation is at least $n/2$. Hence, and this is the point of this section, the LP relaxation behaves in a way completely different from the original integer program.

The complete graph is by no means an isolated case. Dense graphs typically have a maximum independent set much smaller than half of the vertices, and so for such graphs, too, an optimal solution of the LP relaxation tells us almost nothing about the maximum independent set.

It is even known that the size of a maximum independent set cannot be approximated well by any reasonably efficient algorithm whatsoever (provided that some widely believed but unproved assumptions hold, such as $P \neq NP$). This result is from

J. Håstad: Clique is hard to approximate within $n^{1-\varepsilon}$, *Acta Mathematica* 182(1999) 105–142,

and

`http://www.nada.kth.se/~viggo/problemlist/compendium.html`

is a comprehensive website for inapproximability results.

4. Theory of Linear Programming: First Steps

4.1 Equational Form

In the introductory chapter we explained how each linear program can be converted to the form

$$\text{maximize } \mathbf{c}^T\mathbf{x} \text{ subject to } A\mathbf{x} \leq \mathbf{b}.$$

But the simplex method requires a different form, which is usually called the *standard form* in the literature. In this book we introduce a less common, but more descriptive term *equational form*. It looks like this:

Equational form of a linear program:

$$\begin{aligned} \text{Maximize} \quad & \mathbf{c}^T\mathbf{x} \\ \text{subject to} \quad & A\mathbf{x} = \mathbf{b} \\ & \mathbf{x} \geq \mathbf{0}. \end{aligned}$$

As usual, $\mathbf{x}$ is the vector of variables, A is a given $m \times n$ matrix, $\mathbf{c} \in \mathbb{R}^n$, $\mathbf{b} \in \mathbb{R}^m$ are given vectors, and $\mathbf{0}$ is the zero vector, in this case with n components.

The constraints are thus partly equations, and partly inequalities of a very special form $x_j \geq 0$, $j = 1, 2, \ldots, n$, called **nonnegativity constraints**. (Warning: Although we call this form equational, it contains inequalities as well, and these must not be forgotten!)

Let us emphasize that *all* variables in the equational form have to satisfy the nonnegativity constraints.

In problems encountered in practice we often have nonnegativity constraints automatically, since many quantities, such as the amount of consumed cucumber, cannot be negative.

Transformation of an arbitrary linear program to equational form. We illustrate such a transformation for the linear program

$$\begin{aligned} \text{maximize} \quad & 3x_1 - 2x_2 \\ \text{subject to} \quad & 2x_1 - x_2 \leq 4 \\ & x_1 + 3x_2 \geq 5 \\ & x_2 \geq 0. \end{aligned}$$

We proceed as follows:

1. In order to convert the inequality $2x_1 - x_2 \leq 4$ to an equation, we introduce a new variable x_3, together with the nonnegativity constraint $x_3 \geq 0$, and we replace the considered inequality by the equation $2x_1 - x_2 + x_3 = 4$. The auxiliary variable x_3, which won't appear anywhere else in the transformed linear program, represents the difference between the right-hand side and the left-hand side of the inequality. Such an auxiliary variable is called a **slack variable**.

2. For the next inequality $x_1 + 3x_2 \geq 5$ we first multiply by -1, which reverses the direction of the inequality. Then we introduce another slack variable x_4 with the nonnegativity constraint $x_4 \geq 0$, and we replace the inequality by the equation $-x_1 - 3x_2 + x_4 = -5$.

3. We are not finished yet: The variable x_1 in the original linear program is allowed to attain both positive and negative values. We introduce two new, nonnegative, variables y_1 and z_1, $y_1 \geq 0$, $z_1 \geq 0$, and we substitute for x_1 the difference $y_1 - z_1$ everywhere. The variable x_1 itself disappears.

The resulting equational form of our linear program is

$$\begin{aligned}
\text{maximize} \quad & 3y_1 - 3z_1 - 2x_2 \\
\text{subject to} \quad & 2y_1 - 2z_1 - x_2 + x_3 = 4 \\
& -y_1 + z_1 - 3x_2 + x_4 = -5 \\
& y_1 \geq 0, \ z_1 \geq 0, \ x_2 \geq 0, \ x_3 \geq 0, \ x_4 \geq 0.
\end{aligned}$$

So as to comply with the conventions of the equational form in full, we should now rename the variables to $x_1, x_2, \ldots, x_5$.

The presented procedure converts an arbitrary linear program with n variables and m constraints into a linear program in equational form with at most $m + 2n$ variables and m equations (and, of course, nonnegativity constraints for all variables).

Geometry of a linear program in equational form. Let us consider a linear program in equational form:

$$\text{Maximize } \mathbf{c}^T \mathbf{x} \text{ subject to } A\mathbf{x} = \mathbf{b}, \ \mathbf{x} \geq \mathbf{0}.$$

As is derived in linear algebra, the set of all solutions of the system $A\mathbf{x} = \mathbf{b}$ is an affine subspace F of the space $\mathbb{R}^n$. Hence the set of all feasible solutions of the linear program is the intersection of F with the *nonnegative orthant*, which is the set of all points in $\mathbb{R}^n$ with all coordinates nonnegative.[1] The following picture illustrates the geometry of feasible solutions for a linear program with $n = 3$ variables and $m = 1$ equation, namely, the equation $x_1 + x_2 + x_3 = 1$:

---

[1] In the plane ($n = 2$) this set is called the nonnegative *quadrant*, in $\mathbb{R}^3$ it is the nonnegative *octant*, and the name *orthant* is used for an arbitrary dimension.

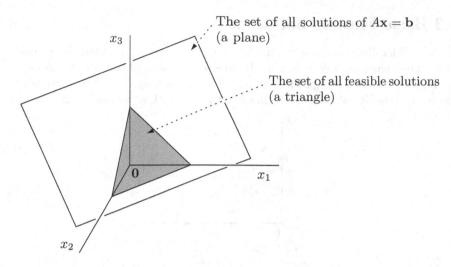

The set of all solutions of $A\mathbf{x} = \mathbf{b}$ (a plane)

The set of all feasible solutions (a triangle)

(In interesting cases we usually have more than 3 variables and no picture can be drawn.)

A preliminary cleanup. Now we will be talking about solutions of the system $A\mathbf{x} = \mathbf{b}$. By this we mean arbitrary real solutions, whose components may be positive, negative, or zero. So this is not the same as feasible solutions of the considered linear program, since a feasible solution has to satisfy $A\mathbf{x} = \mathbf{b}$ *and* have all components nonnegative.

If we change the system $A\mathbf{x} = \mathbf{b}$ by some transformation that preserves the set of solutions, such as a row operation in Gaussian elimination, it influences neither feasible solutions nor optimal solutions of the linear program. This will be amply used in the simplex method.

> **Assumption:** We will consider only linear programs in equational form such that
>
> - the system of equations $A\mathbf{x} = \mathbf{b}$ has at least one solution, and
> - the rows of the matrix A are linearly independent.

As an explanation of this assumption we need to recall a few facts from linear algebra. Checking whether the system $A\mathbf{x} = \mathbf{b}$ has a solution is easy by Gaussian elimination, and if there is no solution, the considered linear program has no feasible solution either, and we can thus disregard it.

If the system $A\mathbf{x} = \mathbf{b}$ has a solution and if some row of A is a linear combination of the other rows, then the corresponding equation is redundant and it can be deleted from the system without changing the set of solutions. We may thus assume that the matrix A has m linearly independent rows and (therefore) rank m.

4.2 Basic Feasible Solutions

Among all feasible solutions of a linear program, a privileged status is granted to so-called basic feasible solutions. In this section we will consider them only for linear programs in equational form. Let us look again at the picture of the set of feasible solutions for a linear program with $n = 3$, $m = 1$:

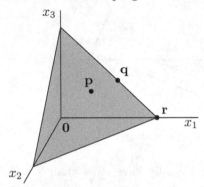

Among the feasible solutions **p**, **q**, and **r** only **r** is basic. Expressed geometrically and very informally, a basic feasible solution is a tip (corner, spike) of the set of feasible solutions. We will formulate this kind of geometric description of a basic feasible solution later (see Theorem 4.4.1).

The definition that we present next turns out to be equivalent, but it looks rather different. It requires that, very roughly speaking, a basic feasible solution have sufficiently many zero components. Before stating it we introduce a new piece of notation.

In this section A is always a matrix with m rows and n columns $(n \geq m)$, of rank m. For a subset $B \subseteq \{1, 2, \ldots, n\}$ we let A_B denote the matrix consisting of the columns of A whose indices belong to B. For instance, for

$$A = \begin{pmatrix} 1 & 5 & 3 & 4 & 6 \\ 0 & 1 & 3 & 5 & 6 \end{pmatrix} \text{ and } B = \{2, 4\} \text{ we have } A_B = \begin{pmatrix} 5 & 4 \\ 1 & 5 \end{pmatrix}.$$

We will use a similar notation for vectors; e.g., for $\mathbf{x} = (3, 5, 7, 9, 11)$ and $B = \{2, 4\}$ we have $\mathbf{x}_B = (5, 9)$.

Now we are ready to state a formal definition.

A **basic feasible solution** of the linear program

$$\text{maximize } \mathbf{c}^T \mathbf{x} \text{ subject to } A\mathbf{x} = \mathbf{b} \text{ and } \mathbf{x} \geq \mathbf{0}$$

is a feasible solution $\mathbf{x} \in \mathbb{R}^n$ for which there exists an m-element set $B \subseteq \{1, 2, \ldots, n\}$ such that

- the (square) matrix A_B is nonsingular, i.e., the columns indexed by B are linearly independent, and
- $x_j = 0$ for all $j \notin B$.

For example, $\mathbf{x} = (0, 2, 0, 1, 0)$ is a basic feasible solution for

$$A = \begin{pmatrix} 1 & 5 & 3 & 4 & 6 \\ 0 & 1 & 3 & 5 & 6 \end{pmatrix}, \quad \mathbf{b} = (14, 7)$$

with $B = \{2, 4\}$.

If such a B is fixed, we call the variables x_j with $j \in B$ the **basic variables**, while the remaining variables are called **nonbasic**. We can thus briefly say that all nonbasic variables are zero in a basic feasible solution.

Let us note that the definition doesn't consider the vector $\mathbf{c}$ at all, and so basic feasible solutions depend solely on A and $\mathbf{b}$.

For some considerations it is convenient to reformulate the definition of a basic feasible solution a little.

4.2.1 Lemma. *A feasible solution $\mathbf{x}$ of a linear program in equational form is basic if and only if the columns of the matrix A_K are linearly independent, where $K = \{j \in \{1, 2, \ldots, n\} : x_j > 0\}$.*

Proof. One of the implications is obvious: If $\mathbf{x}$ is a basic feasible solution and B is the corresponding m-element set as in the definition, then $K \subseteq B$ and thus the columns of the matrix A_K are linearly independent.

Conversely, let $\mathbf{x}$ be feasible and such that the columns of A_K are linearly independent. If $|K| = m$, then we can simply take $B = K$. Otherwise, for $|K| < m$, we extend K to an m-element set B by adding $m - |K|$ more indices so that the columns of A_B are linearly independent. This is a standard fact of linear algebra, which can be verified using the algorithm described next.

We initially set the current B to K, and repeat the following step: If A has a column that is not in the linear span of the columns of A_B, we add the index of such a column to B. As soon as this step is no longer possible, that is, all columns of A are in the linear span of the columns of B, it is easily seen that the columns of A_B constitute a basis of the column space of A. Since A has rank m, we have $|B| = m$ as needed.   □

4.2.2 Proposition. *A basic feasible solution is uniquely determined by the set B. That is, for every m-element set $B \subseteq \{1, 2, \ldots, n\}$ with A_B nonsingular there exists at most one feasible solution $\mathbf{x} \in \mathbb{R}^n$ with $x_j = 0$ for all $j \notin B$.*

Let us stress right away that a single basic feasible solution may be obtained from many different sets B.

Proof of Proposition 4.2.2. For $\mathbf{x}$ to be feasible we must have $A\mathbf{x} = \mathbf{b}$. The left-hand side can be rewritten to $A\mathbf{x} = A_B\mathbf{x}_B + A_N\mathbf{x}_N$, where $N = \{1, 2, \ldots, n\} \setminus B$. For $\mathbf{x}$ to be a basic feasible solution, the vector $\mathbf{x}_N$ of nonbasic variables must equal $\mathbf{0}$, and thus the vector $\mathbf{x}_B$ of basic variables satisfies $A_B\mathbf{x}_B = \mathbf{b}$. And here we use the fact that A_B is a nonsingular square matrix: The system $A_B\mathbf{x}_B = \mathbf{b}$ has exactly one solution $\tilde{\mathbf{x}}_B$. If all components of $\tilde{\mathbf{x}}_B$ are nonnegative, then we have exactly one basic feasible solution for the considered B (we amend $\tilde{\mathbf{x}}_B$ by zeros), and otherwise, we have none.   □

We introduce the following terminology: We call an m-element set $B \subseteq \{1, 2, \ldots, n\}$ with A_B nonsingular a **basis**.[2] If, moreover, B determines a basic feasible solution, or in other words, if the unique solution of the system $A_B \mathbf{x}_B = \mathbf{b}$ is nonnegative, then we call B a **feasible basis**.

The following theorem deals with the existence of optimal solutions, and moreover, it shows that it suffices to look for them solely among basic feasible solutions.

4.2.3 Theorem. *Let us consider a linear program in equational form*

$$\text{maximize } \mathbf{c}^T\mathbf{x} \text{ subject to } A\mathbf{x} = \mathbf{b}, \mathbf{x} \geq \mathbf{0}.$$

(i) *("Optimal solutions may fail to exist only for obvious reasons.") If there is at least one feasible solution and the objective function is bounded from above on the set of all feasible solutions, then there exists an optimal solution.*

(ii) *If an optimal solution exists, then there is a basic feasible solution that is optimal.*

A proof is not necessary for further reading and we defer it to the end of this section. The theorem also follows from the correctness of the simplex method, which will be discussed in the next chapter.

The theorem just stated implies a finite, although entirely impractical, algorithm for solving linear programs in equational form. We consider all m-element subsets $B \subseteq \{1, 2, \ldots, n\}$ one by one, and for each of them we check whether it is a feasible basis, by solving a system of linear equations (we obtain at most one basic feasible solution for each B by Proposition 4.2.2). Then we calculate the maximum of the objective function over all basic feasible solutions found in this way.

Strictly speaking, this algorithm doesn't work if the objective function is unbounded. Formulating a variant of the algorithm that functions properly even in this case, i.e., it reports that the linear program is unbounded, we leave as an exercise. Soon we will discuss the considerably more efficient simplex method, and there we show in detail how to deal with unboundedness.

We have to consider $\binom{n}{m}$ sets B in the above algorithm.[3] For example, for $n = 2m$, the function $\binom{2m}{m}$ grows roughly like 4^m, i.e., exponentially, and this is too much even for moderately large m.

---

[2] This is a shortcut. The index set B itself is not a basis in the sense of linear algebra, of course. Rather the set of columns of the matrix A_B constitutes a basis of the column space of A.

[3] We recall that the binomial coefficient $\binom{n}{m} = \frac{n!}{m!(n-m)!}$ counts the number of m-element subsets of an n-element set.

As we will see in Chapter 5, the simplex method also goes through basic feasible solutions, but in a more clever way. It walks from one to another while improving the value of the objective function all the time, until it reaches an optimal solution.

Let us summarize the main findings of this section.

A linear program *in equational form* has finitely many basic feasible solutions, and if it is feasible and bounded, then at least one of the basic feasible solutions is optimal.

Consequently, any linear program that is feasible and bounded has an optimal solution.

Proof of Theorem 4.2.3. We will use some steps that will reappear in the simplex method in a more elaborate form, and so the present proof is a kind of preparation. We prove the following statement:

> *If the objective function of a linear program in equational form is bounded above, then for every feasible solution $\mathbf{x}_0$ there exists a basic feasible solution $\tilde{\mathbf{x}}$ with the same or larger value of the objective function; i.e., $\mathbf{c}^T \tilde{\mathbf{x}} \geq \mathbf{c}^T \mathbf{x}_0$.*

How does this imply the theorem? If the linear program is feasible and bounded, then according to the statement, for every feasible solution there is a basic feasible solution with the same or larger objective function. Since there are only finitely many basic feasible solutions, some of them have to give the maximum value of the objective function, which means that they are optimal. We thus get both (i) and (ii) at once.

In order to prove the statement, let us consider an arbitrary feasible solution $\mathbf{x}_0$. Among all feasible solutions $\mathbf{x}$ with $\mathbf{c}^T \mathbf{x} \geq \mathbf{c}^T \mathbf{x}_0$ we choose one that has the largest possible number of zero components, and we call it $\tilde{\mathbf{x}}$ (it need not be determined uniquely). We define an index set

$$K = \{j \in \{1, 2, \ldots, n\} : \tilde{x}_j > 0\}.$$

If the columns of the matrix A_K are linearly independent, then $\tilde{\mathbf{x}}$ is a basic feasible solution as in the statement, by Lemma 4.2.1, and we are done.

So let us suppose that the columns of A_K are linearly dependent, which means that there is a nonzero $|K|$-component vector $\mathbf{v}$ such that $A_K \mathbf{v} = \mathbf{0}$. We extend $\mathbf{v}$ by zeros in positions outside K to an n-component vector $\mathbf{w}$ (so $\mathbf{w}_K = \mathbf{v}$ and $A\mathbf{w} = A_K \mathbf{v} = \mathbf{0}$).

Let us assume for a moment that $\mathbf{w}$ satisfies the following two conditions (we will show later why we can assume this):

(i) $\mathbf{c}^T \mathbf{w} \geq 0$.
(ii) There exists $j \in K$ with $w_j < 0$.

For a real number $t \geq 0$ let us consider the vector $\mathbf{x}(t) = \tilde{\mathbf{x}} + t\mathbf{w}$. We show that for some suitable $t_1 > 0$ the vector $\mathbf{x}(t_1)$ is a feasible solution with more zero components than $\tilde{\mathbf{x}}$. At the same time, $\mathbf{c}^T\mathbf{x}(t_1) = \mathbf{c}^T\tilde{\mathbf{x}} + t_1\mathbf{c}^T\mathbf{w} \geq \mathbf{c}^T\mathbf{x}_0 + t_1\mathbf{c}^T\mathbf{w} \geq \mathbf{c}^T\mathbf{x}_0$, and so we get a contradiction to the assumption that $\tilde{\mathbf{x}}$ has the largest possible number of zero components.

We have $A\mathbf{x}(t) = \mathbf{b}$ for all t since $A\mathbf{x}(t) = A\tilde{\mathbf{x}} + tA\mathbf{w} = A\tilde{\mathbf{x}} = \mathbf{b}$, because $\tilde{\mathbf{x}}$ is feasible. Moreover, for $t = 0$ the vector $\mathbf{x}(0) = \tilde{\mathbf{x}}$ has all components from K strictly positive and all other components zero. For the jth component of $\mathbf{x}(t)$ we have $x(t)_j = \tilde{x}_j + tw_j$, and if $w_j < 0$ as in condition (ii), we get $x(t)_j < 0$ for all sufficiently large $t > 0$. If we begin with $t = 0$ and let t grow, then those $x(t)_j$ with $w_j < 0$ are decreasing, and at a certain moment $\tilde{t}$ the first of these decreasing components reaches 0. At this moment, obviously, $\mathbf{x}(\tilde{t})$ still has all components nonnegative, and thus it is feasible, but it has at least one extra zero component compared to $\tilde{\mathbf{x}}$. This, as we have already noted, is a contradiction.

Now what do we do if the vector $\mathbf{w}$ fails to satisfy condition (i) or (ii)? If $\mathbf{c}^T\mathbf{w} = 0$, then (i) holds and (ii) can be recovered by changing the sign of $\mathbf{w}$ (since $\mathbf{w} \neq \mathbf{0}$). So we assume $\mathbf{c}^T\mathbf{w} \neq 0$, and again after a possible sign change we can achieve $\mathbf{c}^T\mathbf{w} > 0$ and thus (i). Now if (ii) fails, we must have $\mathbf{w} \geq 0$. But this means that $\mathbf{x}(t) = \tilde{\mathbf{x}} + t\mathbf{w} \geq 0$ for all $t \geq 0$, and hence all such $\mathbf{x}(t)$ are feasible. The value of the objective function for $\mathbf{x}(t)$ is $\mathbf{c}^T\mathbf{x}(t) = \mathbf{c}^T\tilde{\mathbf{x}} + t\mathbf{c}^T\mathbf{w}$, and it tends to infinity as $t \to \infty$. Hence the linear program is unbounded. This concludes the proof. $\qquad\square$

4.3 ABC of Convexity and Convex Polyhedra

Convexity is one of the fundamental notions in all mathematics, and in the theory of linear programming it is encountered very naturally. Here we recall the definition and present some of the most basic notions and results, which, at the very least, help in gaining a better intuition about linear programming.

On the other hand, linear programming can be presented without these notions, and in concise courses there is usually no time for such material. Accordingly, this section and the next are meant as extending material, and the rest of the book should mostly be accessible without them.

A set $X \subseteq \mathbb{R}^n$ is **convex** if for every two points $\mathbf{x}, \mathbf{y} \in X$ it also contains the segment $\mathbf{xy}$. Expressed differently, for every $\mathbf{x}, \mathbf{y} \in X$ and every $t \in [0, 1]$ we have $t\mathbf{x} + (1-t)\mathbf{y} \in X$.

A word of explanation might be in order: $t\mathbf{x} + (1-t)\mathbf{y}$ is the point on the segment $\mathbf{xy}$ at distance t from $\mathbf{y}$ and distance $1-t$ from $\mathbf{x}$, if we take the length of the segment as unit distance.

Here are a few examples of convex and nonconvex sets in the plane:

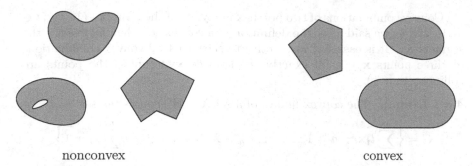

nonconvex   convex

The convex set at the bottom right in this picture, a stadium, is worth re-membering, since often it is a counterexample to statements about convex sets that may look obvious at first sight but are false.

In calculus one works mainly with convex *functions*. Both notions, convex sets and convex functions, are closely related: For instance, a real function $f: \mathbb{R} \to \mathbb{R}$ is convex if and only if its epigraph, i.e., the set $\{(x, y) \in \mathbb{R}^2 : y \geq f(x)\}$, is a convex set in the plane. In general, a function $f : X \to \mathbb{R}$ is called convex if for every $\mathbf{x}, \mathbf{y} \in X$ and every $t \in [0, 1]$ we have

$$f(t\mathbf{x} + (1-t)\mathbf{y}) \leq tf(\mathbf{x}) + (1-t)f(\mathbf{y}).$$

The function is called *strictly convex* if the inequality is strict for all $\mathbf{x} \neq \mathbf{y}$.

Convex hull and convex combinations. It is easily seen that the inter-section of an arbitrary collection of convex sets is again a convex set. This allows us to define the convex hull.

Let $X \subset \mathbb{R}^n$ be a set. The **convex hull** of X is the intersection of all convex sets that contain X. Thus it is the smallest convex set containing X, in the sense that any convex set containing X also contains its convex hull.

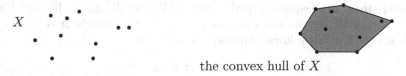

X

the convex hull of X

This is not a very constructive definition. The convex hull can also be described using convex combinations, in a way similar to the description of the linear span of a set of vectors using linear combinations. Let $\mathbf{x}_1, \mathbf{x}_2, \ldots, \mathbf{x}_m$ be points in $\mathbb{R}^n$. Every point of the form

$$t_1\mathbf{x}_1 + t_2\mathbf{x}_2 + \cdots + t_m\mathbf{x}_m, \text{ where } t_1, t_2, \ldots, t_m \geq 0 \text{ and } \sum_{i=1}^{m} t_i = 1,$$

is called a **convex combination** of $\mathbf{x}_1, \mathbf{x}_2, \ldots, \mathbf{x}_m$. A convex combination is thus a particular kind of a linear combination, in which the coefficients are nonnegative and sum to 1.

Convex combinations of two points $\mathbf{x}$ and $\mathbf{y}$ are of the form $t\mathbf{x}+(1-t)\mathbf{y}$, $t \in [0,1]$, and as we said after the definition of a convex set, they fill exactly the segment $\mathbf{xy}$. It is easy but instructive to show that all convex combinations of three points $\mathbf{x}, \mathbf{y}, \mathbf{z}$ fill exactly the triangle $\mathbf{xyz}$ (unless the points are collinear, that is).

4.3.1 Lemma. *The convex hull C of a set $X \subseteq \mathbb{R}^n$ equals the set*

$$\tilde{C} = \left\{ \sum_{i=1}^m t_i \mathbf{x}_i : m \geq 1, \mathbf{x}_1, \ldots, \mathbf{x}_m \in X, t_1, \ldots, t_m \geq 0, \sum_{i=1}^m t_i = 1 \right\}$$

of all convex combinations of finitely many points of X.

Proof. First we prove by induction on m that each convex combination has to lie in the convex hull C. For $m = 1$ it is obvious and for $m = 2$ it follows directly from the convexity of C.

Let $m \geq 3$ and let $\mathbf{x} = t_1 \mathbf{x}_1 + \cdots + t_m \mathbf{x}_m$ be a convex combination of points of X. If $t_m = 1$, then we have $\mathbf{x} = \mathbf{x}_m \in C$. For $t_m < 1$ let us put $t_i' = t_i/(1 - t_m)$, $i = 1, 2, \ldots, m - 1$. Then $\mathbf{x}' = t_1' \mathbf{x}_1 + \cdots + t_{m-1}' \mathbf{x}_{m-1}$ is a convex combination of the points $\mathbf{x}_1, \ldots, \mathbf{x}_{m-1}$ (the t_i' sum to 1), and by the inductive hypothesis $\mathbf{x}' \in C$. So $\mathbf{x} = (1 - t_m)\mathbf{x}' + t_m \mathbf{x}_m$ is a convex combination of two points of the (convex) set C and as such it also lies in C.

We have thus proved $\tilde{C} \subseteq C$. For the reverse inclusion it suffices to prove that $\tilde{C}$ is convex, that is, to verify that whenever $\mathbf{x}, \mathbf{y} \in \tilde{C}$ are two convex combinations and $t \in (0,1)$, then $t\mathbf{x}+(1-t)\mathbf{y}$ is again a convex combination. This is straightforward and we take the liberty of omitting further details. $\square$

Convex sets encountered in the theory of linear programming are of a special type and they are called convex polyhedra.

Hyperplanes, half-spaces, polyhedra. We recall that a **hyperplane** in $\mathbb{R}^n$ is an affine subspace of dimension $n-1$. In other words, it is the set of all solutions of a single linear equation of the form

$$a_1 x_1 + a_2 x_2 + \cdots + a_n x_n = b,$$

where $a_1, a_2, \ldots, a_n$ are not all 0. Hyperplanes in $\mathbb{R}^2$ are lines and hyperplanes in $\mathbb{R}^3$ are ordinary planes.

A hyperplane divides $\mathbb{R}^n$ into two half-spaces and it constitutes their common boundary. For the hyperplane with equation $a_1 x_1 + a_2 x_2 + \cdots + a_n x_n = b$, the two half-spaces have the following analytic expression:

$$\left\{ \mathbf{x} \in \mathbb{R}^n : a_1 x_1 + a_2 x_2 + \cdots + a_n x_n \leq b \right\}$$

and

$$\left\{ \mathbf{x} \in \mathbb{R}^n : a_1 x_1 + a_2 x_2 + \cdots + a_n x_n \geq b \right\}.$$

More exactly, these are **closed half-spaces** that contain their boundary.

> A **convex polyhedron** is an intersection of finitely many closed half-spaces in $\mathbb{R}^n$.

A half-space is obviously convex, and hence an intersection of half-spaces is convex as well. Thus convex polyhedra bear the attribute convex by right.

A disk in the plane is a convex set, but it is not a convex polyhedron (because, roughly speaking, a convex polyhedron has to be "edgy"... but try proving this formally).

A half-space is the set of all solutions of a single linear inequality (with at least one nonzero coefficient of some variable x_j). The set of all solutions of a system of finitely many linear inequalities, a.k.a. the set of all feasible solutions of a linear program, is geometrically the intersection of finitely many half-spaces, alias a convex polyhedron. (We should perhaps also mention that a hyperplane is the intersection of two half-spaces, and so the constraints can be both inequalities and equations.)

Let us note that a convex polyhedron can be unbounded, since, for example, a single half-space is also a convex polyhedron. A bounded convex polyhedron, i.e. one that can be placed inside some large enough ball, is called a *convex polytope*.

The **dimension** of a convex polyhedron $P \subseteq \mathbb{R}^n$ is the smallest dimension of an affine subspace containing P. Equivalently, it is the largest d for which P contains points $\mathbf{x}_0, \mathbf{x}_1, \ldots, \mathbf{x}_d$ such that the d-tuple of vectors $(\mathbf{x}_1 - \mathbf{x}_0, \mathbf{x}_2 - \mathbf{x}_0, \ldots, \mathbf{x}_d - \mathbf{x}_0)$ is linearly independent.

The empty set is also a convex polyhedron, and its dimension is usually defined as -1.

All convex polygons in the plane are two-dimensional convex polyhedra. Several types of three-dimensional convex polyhedra are taught at high schools and decorate mathematical cabinets, such as cubes, boxes, pyramids, or even regular dodecahedra, which can also be met as desktop calendars. Simple examples of convex polyhedra of an arbitrary dimension n are:

- The n-dimensional **cube** $[-1,1]^n$, which can be written as the intersection of $2n$ half-spaces (which ones?):

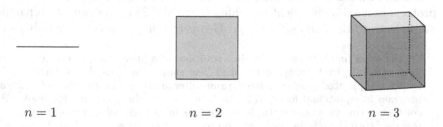

$n = 1$ $n = 2$ $n = 3$

- the n-dimensional **crosspolytope** $\{\mathbf{x} \in \mathbb{R}^n : |x_1| + |x_2| + \cdots + |x_n| \leq 1\}$:

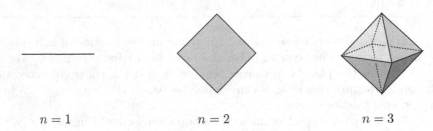

$$n = 1 \qquad\qquad n = 2 \qquad\qquad n = 3$$

For $n = 3$ we get the regular octahedron. For expressing the n-dimensional crosspolytope as an intersection of half-spaces we need 2^n half-spaces (can you find them?).

- The regular n-dimensional **simplex**

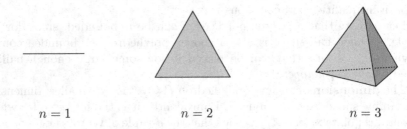

$$n = 1 \qquad\qquad n = 2 \qquad\qquad n = 3$$

can be defined in a quite simple and nice way as a subset of $\mathbb{R}^{n+1}$:

$$\{\mathbf{x} \in \mathbb{R}^{n+1} : x_1, x_2, \ldots, x_{n+1} \geq 0,\ x_1 + x_2 + \cdots + x_{n+1} = 1\}.$$

We note that this is exactly the set of all feasible solutions of the linear program with the single equation $x_1 + x_2 + \cdots + x_{n+1} = 1$ and non-negativity constraints;[4] see the picture in Section 4.1. In general, any n-dimensional convex polytope bounded by $n+1$ hyperplanes is called a simplex.

Many interesting examples of convex polyhedra are obtained as sets of feasible solutions of natural linear programs. For example, the LP relaxation of the problem of maximum-weight matching (Section 3.2) for a complete bipartite graph leads to the *Birkhoff polytope*. Geometric properties of such polyhedra

---

[4] On the other hand, the set of feasible solutions of a linear program in equational form certainly isn't always a simplex! The simplex method is so named for a rather complicated reason, related to an alternative geometric view of a linear program in equational form, different from the one discussed in this book. According to this view, the m-tuple of numbers in the jth *column* of the matrix A together with the number c_j is interpreted as a point in $\mathbb{R}^{m+1}$. Then the simplex method can be interpreted as a walk through certain simplices with vertices at these points. It was this view that gave Dantzig faith in the simplex method and convinced him that it made sense to study it.

are often related to properties of combinatorial objects and to solutions of combinatorial optimization problems in an interesting way. A nice book about convex polyhedra is

G. M. Ziegler: *Lectures on Polytopes*, Springer-Verlag, Heidelberg, 1994 (corrected 2nd edition 1998).

The book

B. Grünbaum: *Convex Polytopes*, second edition prepared by Volker Kaibel, Victor Klee, and Günter Ziegler, Springer-Verlag, Heidelberg, 2003

is a new edition of a 1967 classics, with extensive updates on the material covered in the original book.

4.4 Vertices and Basic Feasible Solutions

A vertex of a convex polyhedron can be thought of as a "tip" or "spike." For instance, a three-dimensional cube has 8 vertices, and a regular octahedron has 6 vertices.

Mathematically, a vertex is defined as a point where some linear function attains a unique maximum. Thus a point $\mathbf{v}$ is called a **vertex** of a convex polyhedron $P \subset \mathbb{R}^n$ if $\mathbf{v} \in P$ and there exists a nonzero vector $\mathbf{c} \in \mathbb{R}^n$ such that $\mathbf{c}^T \mathbf{v} > \mathbf{c}^T \mathbf{y}$ for all $\mathbf{y} \in P \setminus \{\mathbf{v}\}$. Geometrically it means that the hyperplane $\{\mathbf{x} \in \mathbb{R}^n : \mathbf{c}^T \mathbf{x} = \mathbf{c}^T \mathbf{v}\}$ touches the polyhedron P exactly at $\mathbf{v}$.

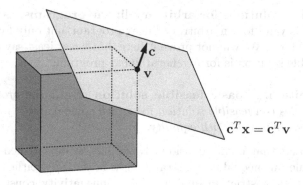

Three-dimensional polyhedra have not only vertices, but also edges and faces. A general polyhedron $P \subseteq \mathbb{R}^n$ of dimension n can have vertices, edges, 2-dimensional faces, 3-dimensional faces, up to $(n-1)$-dimensional faces. They are defined as follows: A subset $F \subseteq P$ is a **k-dimensional face** of a convex polyhedron P if F has dimension k and there exist a nonzero vector $\mathbf{c} \in \mathbb{R}^n$ and a number $z \in \mathbb{R}$ such that

$\mathbf{c}^T\mathbf{x} = z$ for all $\mathbf{x} \in F$ and $\mathbf{c}^T\mathbf{x} < z$ for all $\mathbf{x} \in P \setminus F$. In other words, there exists a hyperplane that touches P exactly at F. Since such an F is the intersection of a hyperplane with a convex polyhedron, it is a convex polyhedron itself, and its dimension is thus well defined. An **edge** is a 1-dimensional face and a vertex is a 0-dimensional face.

Now we prove that vertices of a convex polyhedron and basic feasible solutions of a linear program are the same concept.

4.4.1 Theorem. *Let P be the set of all feasible solutions of a linear program in equational form (so P is a convex polyhedron). Then the following two conditions for a point $\mathbf{v} \in P$ are equivalent:*

(i) $\mathbf{v}$ *is a vertex of the polyhedron P.*
(ii) $\mathbf{v}$ *is a basic feasible solution of the linear program.*

Proof. The implication (i)$\Rightarrow$(ii) follows immediately from Theorem 4.2.3 (with $\mathbf{c}$ being the vector defining $\mathbf{v}$). It remains to prove (ii)$\Rightarrow$(i).

Let us consider a basic feasible solution $\mathbf{v}$ with a feasible basis B, and let us define a vector $\tilde{\mathbf{c}} \in \mathbb{R}^n$ by $\tilde{c}_j = 0$ for $j \in B$ and $\tilde{c}_j = -1$ otherwise. We have $\tilde{\mathbf{c}}^T\mathbf{v} = 0$, and $\tilde{\mathbf{c}}^T\mathbf{x} \leq 0$ for any $\mathbf{x} \geq \mathbf{0}$, and hence $\mathbf{v}$ maximizes the objective function $\tilde{\mathbf{c}}^T\mathbf{x}$. Moreover, $\tilde{\mathbf{c}}^T\mathbf{x} < 0$ whenever $\mathbf{x}$ has a nonzero component outside B. But by Proposition 4.2.2, $\mathbf{v}$ is the *only* feasible solution with all nonzero components in B, and therefore $\mathbf{v}$ is the only point of P maximizing $\tilde{\mathbf{c}}^T\mathbf{x}$. $\qquad\square$

Basic feasible solutions for arbitrary linear programs. A similar theorem is valid for an arbitrary linear program, not only for one in equational form. We will not prove it here, but we at least say what a basic feasible solution is for a general linear program:

4.4.2 Definition. *A basic feasible solution of a linear program with n variables is a feasible solution for which some n linearly independent constraints hold with equality.*

A constraint that is an equation always has to be satisfied with equality, while an inequality constraint may be satisfied either with equality or with a strict inequality. The nonnegativity constraints satisfied with equality are also counted. The linear independence of constraints means that the vectors of the coefficients of the variables are linearly independent. For example, for $n = 4$, the constraint $3x_1 + 5x_3 - 7x_4 \leq 10$ has the corresponding vector $(3, 0, 5, -7)$.

As is known from linear algebra, a system of n linearly independent linear equations in n variables has exactly one solution. Hence, if $\mathbf{x}$ is a basic feasible solution and it satisfies some n linearly independent constraints with equality, then it is the only point in $\mathbb{R}^n$ that satisfies these n constraints with equality. Geometrically speaking, the constraints satisfied with equality determine hyperplanes, $\mathbf{x}$ lies on some n of them, and these n hyperplanes meet in a single point.

The definition of a basic feasible solution for the equational form looks quite different, but in fact, it is a special case of the new definition, as we now indicate. For a linear program in equational form we have m linearly independent equations always satisfied with equality, and so it remains to satisfy with equality some $n - m$ of the nonnegativity constraints, and these must be linearly independent with the equations. The coefficient vector of the nonnegativity constraint $x_j \geq 0$ is $\mathbf{e}_j$, with 1 at position j and with zeros elsewhere. If $\mathbf{x}$ is a basic feasible solution according to the new definition, then there exists a set $N \subseteq \{1, 2, \ldots, n\}$ of size $n - m$ such that $x_j = 0$ for all $j \in N$ and the rows of the matrix A together with the vectors $(\mathbf{e}_j : j \in N)$ constitute a linearly independent collection. This happens exactly if the matrix A_B has linearly independent rows, where $B = \{1, 2, \ldots, n\} \setminus N$, and we are back at the definition of a basic feasible solution for the equational form.

For a general linear program none of the optimal solutions have to be basic, as is illustrated by the linear program

$$\text{maximize } x_1 + x_2 \text{ subject to } x_1 + x_2 \leq 1.$$

This contrasts with the situation for the equational form (cf. Theorem 4.2.3) and it is one of the advantages of the equational form.

Vertices and extremal points. The intuitive notion of a "tip" of a convex set can be viewed mathematically in at least two ways. One of them is captured by the above definition of a vertex of a convex polyhedron: A tip is a point for which some linear function attains a unique maximum. The other one leads to a definition talking about points that cannot be "generated by segments." These are called **extremal points**; thus a point $\mathbf{x}$ is an extremal point of a convex set $C \subseteq \mathbb{R}^n$ if $\mathbf{x} \in C$ and there are no two points $\mathbf{y}, \mathbf{z} \in C$ different from $\mathbf{x}$ such that $\mathbf{x}$ lies on the segment $\mathbf{y}\mathbf{z}$.

For a convex polyhedron it is not difficult to show that the extremal points are exactly the vertices. Hence we have yet another equivalent description of a basic feasible solution.

A convex polytope is the convex hull of its vertices. A general convex polyhedron need not have any vertices at all—consider a half-space. However, a convex *polytope* P, i.e., a bounded convex polyhedron, always has vertices, and even more is true: P equals the convex hull of the set of its vertices. This may look intuitively obvious from examples in dimensions 2 and 3, but a proof is nontrivial (Ziegler's book cited in the previous section calls this the "Main The-

orem" of polytope theory). Consequently, every convex polytope can
be represented either as the intersection of finitely many half-spaces
or as the convex hull of finitely many points.

5. The Simplex Method

In this chapter we explain the simplex method for solving linear programs. We will make use of the terms *equational form* and *basic feasible solution* from the previous chapter.

Gaussian elimination in linear algebra has a fundamental theoretical and didactic significance, as a starting point for further developments. But in practice it has mostly been replaced by more complicated and more efficient algorithms. Similarly, the basic version of the simplex method that we discuss here is not commonly used for solving linear programs in practice. We do not put emphasis on the most efficient possible organization of the computations, but rather we concentrate on the main ideas.

5.1 An Introductory Example

We will first show the simplex method in action on a small concrete example, namely, on the following linear program:

$$
\begin{array}{lrcl}
\text{Maximize} & x_1 + x_2 & & \\
\text{subject to} & -x_1 + x_2 & \leq & 1 \\
& x_1 & \leq & 3 \\
& x_2 & \leq & 2 \\
& x_1, x_2 & \geq & 0.
\end{array}
\tag{5.1}
$$

We intentionally do not take a linear program in equational form: The variables are nonnegative, but the inequalities have to be replaced by equations, by introducing slack variables. The equational form is

$$
\begin{array}{lrcll}
\text{maximize} & x_1 + x_2 & & & \\
\text{subject to} & -x_1 + x_2 + x_3 & & & = 1 \\
& x_1 & + x_4 & & = 3 \\
& x_2 & & + x_5 & = 2 \\
& x_1, x_2, \ldots, x_5 \geq 0, & & &
\end{array}
$$

with the matrix

$$
A = \begin{pmatrix} -1 & 1 & 1 & 0 & 0 \\ 1 & 0 & 0 & 1 & 0 \\ 0 & 1 & 0 & 0 & 1 \end{pmatrix}.
$$

In the simplex method we first express each linear program in the form of a *simplex tableau*. In our case we begin with the tableau

$$
\begin{aligned}
x_3 &= 1 + x_1 - x_2 \\
x_4 &= 3 - x_1 \\
x_5 &= 2 \qquad\quad - x_2 \\
\hline
z &= \qquad\quad x_1 + x_2
\end{aligned}
$$

The first three rows consist of the equations of the linear program, in which the slack variables have been carried over to the left-hand side and the remaining terms are on the right-hand side. The last row, separated by a line, contains a new variable z, which expresses the objective function.

Each simplex tableau is associated with a certain basic feasible solution. In our case we substitute 0 for the variables x_1 and x_2 from the right-hand side, and without calculation we see that $x_3 = 1, x_4 = 3, x_5 = 2$. This feasible solution is indeed basic with $B = \{3, 4, 5\}$; we note that A_B is the identity matrix. The variables x_3, x_4, x_5 from the left-hand side are basic and the variables x_1, x_2 from the right-hand side are nonbasic. The value of the objective function $z = 0$ corresponding to this basic feasible solution can be read off from the last row of the tableau.

From the initial simplex tableau we will construct a sequence of tableaus of a similar form, by gradually rewriting them according to certain rules. Each tableau will contain the *same* information about the linear program, only written differently. The procedure terminates with a tableau that represents the information so that the desired optimal solution can be read off directly.

Let us go to the first step. We try to increase the value of the objective function by increasing one of the nonbasic variables x_1 or x_2. In the above tableau we observe that increasing the value of x_1 (i.e. making x_1 positive) increases the value of z. The same is true for x_2, because both variables have positive coefficients in the z-row of the tableau. We can choose either x_1 or x_2; let us decide (arbitrarily) for x_2. We will increase it, while x_1 will stay 0.

By how much can we increase x_2? If we want to maintain feasibility, we have to be careful not to let any of the basic variables x_3, x_4, x_5 go below zero. This means that the equations determining x_3, x_4, x_5 may limit the increment of x_2. Let us consider the first equation

$$x_3 = 1 + x_1 - x_2.$$

Together with the implicit constraint $x_3 \geq 0$ it lets us increase x_2 up to the value $x_2 = 1$ (while keeping $x_1 = 0$). The second equation

$$x_4 = 3 - x_1$$

does not limit the increment of x_2 at all, and the third equation

$$x_5 = 2 - x_2$$

allows for an increase of x_2 up to $x_2 = 2$ before x_5 gets negative. The most stringent restriction thus follows from the first equation.

We increase x_2 as much as we can, obtaining $x_2 = 1$ and $x_3 = 0$. From the remaining equations of the tableau we get the values of the other variables:

$$x_4 = 3 - x_1 = 3$$
$$x_5 = 2 - x_2 = 1.$$

In this new feasible solution x_3 became zero and x_2 nonzero. Quite naturally we thus transfer x_3 to the right-hand side, where the nonbasic variables live, and x_2 to the left-hand side, where the basic variables reside. We do it by means of the most stringent equation $x_3 = 1 + x_1 - x_2$, from which we express

$$x_2 = 1 + x_1 - x_3.$$

We substitute the right-hand side for x_2 into the remaining equations, and we arrive at a new tableau:

$$
\begin{aligned}
x_2 &= 1 + x_1 - x_3 \\
x_4 &= 3 - x_1 \\
x_5 &= 1 - x_1 + x_3 \\
\hline
z &= 1 + 2x_1 - x_3
\end{aligned}
$$

Here $B = \{2,4,5\}$, which corresponds to the basic feasible solution $\mathbf{x} = (0,1,0,3,1)$ with the value of the objective function $z = 1$.

This process of rewriting one simplex tableau into another is called a **pivot step**. In each pivot step some nonbasic variable, in our case x_2, *enters* the basis, while some basic variable, in our case x_3, *leaves* the basis.

In the new tableau we can further increase the value of the objective function by increasing x_1, while increasing x_3 would lead to a smaller z-value. The first equation does not restrict the increment of x_1 in any way, from the second one we get $x_1 \leq 3$, and from the third one $x_1 \leq 1$, so the strictest limitation is implied by the third equation. Similarly as in the previous step, we express x_1 from it and we substitute this expression into the remaining equations. Thereby x_1 enters the basis and moves to the left-hand side, and x_5 leaves the basis and migrates to the right-hand side. The tableau we obtain is

$$
\begin{aligned}
x_1 &= 1 + x_3 - x_5 \\
x_2 &= 2 - x_5 \\
x_4 &= 2 - x_3 + x_5 \\
\hline
z &= 3 + x_3 - 2x_5
\end{aligned}
$$

with $B = \{1,2,4\}$, basic feasible solution $\mathbf{x} = (1,2,0,2,0)$, and $z = 3$. After one more pivot step, in which x_3 enters the basis and x_4 leaves it, we arrive at the tableau

$$
\begin{aligned}
x_1 &= 3 - x_4 \\
x_2 &= 2 - x_5 \\
x_3 &= 2 - x_4 + x_5 \\
\hline
z &= 5 - x_4 - x_5
\end{aligned}
$$

with basis $\{1, 2, 3\}$, basic feasible solution $\mathbf{x} = (3, 2, 2, 0, 0)$, and $z = 5$. In this tableau, no nonbasic variable can be increased without making the objective function value smaller, so we are stuck. Luckily, this also means that we have already found an optimal solution! Why?

Let us consider an *arbitrary* feasible solution $\tilde{\mathbf{x}} = (\tilde{x}_1, \ldots, \tilde{x}_5)$ of our linear program, with the objective function attaining some value $\tilde{z}$. Now $\tilde{\mathbf{x}}$ and $\tilde{z}$ satisfy all equations in the final tableau, which was obtained from the original equations of the linear program by equivalent transformations. Hence we necessarily have

$$\tilde{z} = 5 - \tilde{x}_4 - \tilde{x}_5.$$

Together with the nonnegativity constraints $\tilde{x}_4, \tilde{x}_5 \geq 0$ this implies $\tilde{z} \leq 5$. The tableau even delivers a proof that $\mathbf{x} = (3, 2, 2, 0, 0)$ is the *only* optimal solution: If $z = 5$, then $x_4 = x_5 = 0$, and this determines the values of the remaining variables uniquely.

A geometric illustration. For each feasible solution (x_1, x_2) of the original linear program (5.1) with inequalities we have exactly one corresponding feasible solution $(x_1, x_2, \ldots, x_5)$ of the modified linear program in equational form, and conversely. The sets of feasible solutions are isomorphic in a suitable sense, and we can thus follow the progress of the simplex method narrated above in a planar picture for the original linear program (5.1):

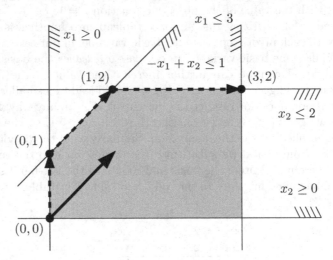

We can see the simplex method moving along the edges from one feasible solution to another, while the value of the objective function grows until it reaches the optimum. In the example we could also take a shorter route if we decided to increase x_1 instead of x_2 in the first step.

Potential troubles. In our modest example the simplex method has run smoothly without any problems. In general we must deal with several complications. We will demonstrate them on examples in the next sections.

5.2 Exception Handling: Unboundedness

What happens in the simplex method for an unbounded linear program? We will show it on the example

$$\begin{array}{ll} \text{maximize} & x_1 \\ \text{subject to} & x_1 - x_2 \leq 1 \\ & -x_1 + x_2 \leq 2 \\ & x_1, x_2 \geq 0 \end{array}$$

illustrated in the picture below:

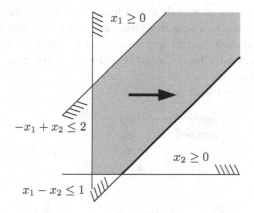

After the usual transformation to equational form by introducing slack variables x_3, x_4, we can use these variables as a feasible basis and we obtain the initial simplex tableau

$$\begin{array}{rcl} x_3 &=& 1 - x_1 + x_2 \\ x_4 &=& 2 + x_1 - x_2 \\ \hline z &=& x_1 \end{array}$$

After the first pivot step with entering variable x_1 and leaving variable x_3 the next tableau is

$$\begin{array}{rcl} x_1 &=& 1 + x_2 - x_3 \\ x_4 &=& 3 \quad\quad - x_3 \\ \hline z &=& 1 + x_2 - x_3 \end{array}$$

If we now try to introduce x_2 into the basis, we discover that none of the equations in the tableau restrict its increase in any way. We can thus take x_2 arbitrarily large, and we also get z arbitrarily large—the linear program is unbounded.

Let us analyze this situation in more detail. From the tableau one can see that for an arbitrarily large number $t \geq 0$ we obtain a feasible solution by setting $x_2 = t$, $x_3 = 0$, $x_1 = 1+t$, and $x_4 = 3$, with the value of the objective function $z = 1 + t$. In other words, the semi-infinite ray

$$\{(1, 0, 0, 3) + t(1, 1, 0, 0) : t \geq 0\}$$

is contained in the set of feasible solutions. It "witnesses" the unboundedness of the linear program, since the objective function attains arbitrarily large values on it. The corresponding semi-infinite ray for the original two-dimensional linear program is drawn thick in the picture above.

A similar ray is the output of the simplex method for all unbounded linear programs.

5.3 Exception Handling: Degeneracy

While we can make some nonbasic variable arbitrarily large in the unbounded case, the other extreme happens in a situation called a degeneracy: The equations in a tableau do not permit any increment of the selected nonbasic variable, and it may actually be impossible to increase the objective function z in a single pivot step.

Let us consider the linear program

$$
\begin{aligned}
\text{maximize} \quad & x_2 \\
\text{subject to} \quad & -x_1 + x_2 \leq 0 \\
& x_1 \qquad\;\; \leq 2 \\
& x_1, x_2 \geq 0.
\end{aligned}
\tag{5.2}
$$

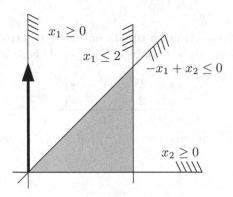

In the usual way we convert it to equational form and construct the initial tableau

$$
\begin{aligned}
x_3 &= \quad\;\; x_1 - x_2 \\
\underline{x_4 &= 2 - x_1 \qquad\quad} \\
z &= \qquad\quad\; x_2
\end{aligned}
$$

The only candidate for entering the basis is x_2, but the first row of the tableau shows that its value cannot be increased without making x_3 negative. Unfortunately, the impossibility of making progress in this case does not imply optimality, so we have to perform a degenerate pivot step, i.e., one with zero progress in the objective function. In our example, bringing x_2 into the basis (with x_3 leaving) results in another tableau with the same basic feasible solution $(0, 0, 0, 2)$:

$$
\begin{array}{rl}
x_2 = & x_1 - x_3 \\
x_4 = 2 - & x_1 \\
\hline
z = & x_1 - x_3
\end{array}
$$

Nevertheless, the situation has improved. The nonbasic variable x_1 can now be increased, and by entering it into the basis (replacing x_4) we already obtain the final tableau

$$
\begin{array}{rl}
x_1 = 2 & - x_4 \\
x_2 = 2 - x_3 & - x_4 \\
\hline
z = 2 - x_3 & - x_4
\end{array}
$$

with an optimal solution $\mathbf{x} = (2, 2, 0, 0)$.

A situation that forces a degenerate pivot step may occur only for a linear program in which several feasible bases correspond to a single basic feasible solution. Such linear programs are called **degenerate**.

It is easily seen that in order that a single basic feasible solution be obtained from several bases, some of the *basic* variables have to be zero.

In this example, after one degenerate pivot step we could again make progress. In general, there might be longer runs of degenerate pivot steps. It may even happen that some tableau is repeated in a sequence of degenerate pivot steps, and so the algorithm might pass through an infinite sequence of tableaus without any progress. This phenomenon is called **cycling**. An example of a linear program for which the simplex method may cycle can be found in Chvátal's textbook cited in Chapter 9 (the smallest possible example has 6 variables and 3 equations), and we will not present it here.

If the simplex method doesn't cycle, then it necessarily finishes in a finite number of steps. This is because there are only finitely many possible simplex tableaus for any given linear program, namely at most $\binom{n}{m}$, which we will prove in Section 5.5.

How can cycling be prevented? This is a nontrivial issue and it will be discussed in Section 5.8.

5.4 Exception Handling: Infeasibility

In order that the simplex method be able to start at all, we need a feasible basis. In examples discussed up until now we got a feasible basis more or less for free. It works this way for all linear programs of the form

$$\text{maximize } \mathbf{c}^T \mathbf{x} \text{ subject to } A\mathbf{x} \leq \mathbf{b} \text{ and } \mathbf{x} \geq \mathbf{0}$$

with $\mathbf{b} \geq \mathbf{0}$. Indeed, the indices of the slack variables introduced in the transformation to equational form can serve as a feasible basis.

However, in general, finding any feasible solution of a linear program is equally as difficult as finding an optimal solution (see the remark in Section 1.3). But computing the initial feasible basis can be done by the simplex method itself, if we apply it to a suitable auxiliary problem.

Let us consider the linear program in equational form

$$\begin{array}{ll} \text{maximize} & x_1 + 2x_2 \\ \text{subject to} & x_1 + 3x_2 + x_3 = 4 \\ & \phantom{x_1 + {}} 2x_2 + x_3 = 2 \\ & x_1, x_2, x_3 \geq 0. \end{array}$$

Let us try to produce a feasible solution starting with $(x_1, x_2, x_3) = (0, 0, 0)$. This vector is nonnegative, but of course it is not feasible, since it does not satisfy the equations of the linear program. We introduce auxiliary variables x_4 and x_5 as "corrections" of infeasibility: $x_4 = 4 - x_1 - 3x_2 - x_3$ expresses by how much the original variables x_1, x_2, x_3 fail to satisfy the first equation, and $x_5 = 2 - 2x_2 - x_3$ plays a similar role for the second equation. If we managed to find nonnegative values of x_1, x_2, x_3 for which both of these corrections come out as zeros, we would have a feasible solution of the considered linear program.

The task of finding nonnegative x_1, x_2, x_3 with zero corrections can be captured by a linear program:

$$\begin{array}{llll} \text{Maximize} & & - x_4 - x_5 \\ \text{subject to} & x_1 + 3x_2 + x_3 + x_4 & = 4 \\ & \phantom{x_1 + {}} 2x_2 + x_3 \phantom{{} + x_4} + x_5 & = 2 \\ & x_1, x_2, \ldots, x_5 \geq 0. \end{array}$$

The optimal value of the objective function $-x_4 - x_5$ is 0 exactly if there exist values of x_1, x_2, x_3 with zero corrections, i.e., a feasible solution of the original linear program.

This is the right auxiliary linear program. The variables x_4 and x_5 form a feasible basis, with the basic feasible solution $(0, 0, 0, 4, 2)$. (Here we use that the right-hand sides, 4 and 2, are nonnegative, but since we deal with *equations*, this can always be achieved by sign changes.) Once we express the objective function using the nonbasic variables, that is, in the form $z = -6 + x_1 + 5x_2 + 2x_3$, we can start the simplex method on the auxiliary linear program.

The auxiliary linear program is surely bounded, since the objective function cannot be positive. The simplex method thus computes a basic feasible solution that is optimal.

As training the reader can check that if we let x_1 enter the basis in the first pivot step and x_3 in the second, the final simplex tableau comes out as

$$\begin{array}{rl} x_1 =& 2 - x_2 - x_4 + x_5 \\ x_3 =& 2 - 2x_2 \phantom{{} - x_4} - x_5 \\ \hline z =& - x_4 - x_5. \end{array}$$

The corresponding optimal solution $(2, 0, 2, 0, 0)$ yields a basic feasible solution of the original linear program: $(x_1, x_2, x_3) = (2, 0, 2)$. The initial simplex tableau for the original linear program can even be obtained from the final tableau of the auxiliary linear program, by leaving out the columns of the auxiliary variables x_4 and x_5,[1] and by changing the objective function back to the original one, expressed in terms of the nonbasic variables:

$$
\begin{aligned}
x_1 &= 2 - x_2 \\
x_3 &= 2 - 2x_2 \\
\hline
z &= 2 + x_2
\end{aligned}
$$

Starting from this tableau, a single pivot step already reaches the optimum.

5.5 Simplex Tableaus in General

In this section and the next one we formulate in general, and mostly with proofs, what has previously been explained on examples.

Let us consider a general linear program in equational form

$$\text{maximize } \mathbf{c}^T \mathbf{x} \text{ subject to } A\mathbf{x} = \mathbf{b} \text{ and } \mathbf{x} \geq \mathbf{0}.$$

The simplex method applied to it computes a sequence of simplex tableaus. Each of them corresponds to a feasible basis B and it determines a basic feasible solution, as we will soon verify. (Let us recall that a feasible basis is an m-element set $B \subseteq \{1, 2, \ldots, n\}$ such that the matrix A_B is nonsingular and the (unique) solution of the system $A_B \mathbf{x}_B = \mathbf{b}$ is nonnegative.)

Formally, we will define a simplex tableau as a certain system of linear equations of a special form, in which the basic variables and the variable z, representing the value of the objective function, stand on the left-hand side and they are expressed in terms of the nonbasic variables.

A **simplex tableau** $\mathcal{T}(B)$ determined by a feasible basis B is a system of $m+1$ linear equations in variables $x_1, x_2, \ldots, x_n$, and z that has the *same set of solutions* as the system $A\mathbf{x} = \mathbf{b}$, $z = \mathbf{c}^T\mathbf{x}$, and in matrix notation looks as follows:

$$
\begin{aligned}
\mathbf{x}_B &= \mathbf{p} + Q\mathbf{x}_N \\
\hline
z &= z_0 + \mathbf{r}^T\mathbf{x}_N
\end{aligned}
$$

where $\mathbf{x}_B$ is the vector of the basic variables, $N = \{1, 2, \ldots, n\} \setminus B$, $\mathbf{x}_N$ is the vector of nonbasic variables, $\mathbf{p} \in \mathbb{R}^m$, $\mathbf{r} \in \mathbb{R}^{n-m}$, Q is an $m \times (n-m)$ matrix, and $z_0 \in \mathbb{R}$.

---

[1] It may happen that some auxiliary variables are zero but still basic in the final tableau of the auxiliary program, and so they cannot simply be left out. Section 5.6 discusses this (easy) issue.

The basic feasible solution corresponding to this tableau can be read off immediately: It is obtained by substituting $\mathbf{x}_N = \mathbf{0}$; that is, we have $\mathbf{x}_B = \mathbf{p}$. From the feasibility of the basis B we see that $\mathbf{p} \geq \mathbf{0}$. The objective function for this basic feasible solution has value $z_0 + \mathbf{r}^T \mathbf{0} = z_0$.

The values of $\mathbf{p}, Q, \mathbf{r}, z_0$ can easily be expressed using B and $A, \mathbf{b}, \mathbf{c}$:

5.5.1 Lemma. *For each feasible basis B there exists exactly one simplex tableau, and it is given by*

$$Q = -A_B^{-1} A_N, \quad \mathbf{p} = A_B^{-1} \mathbf{b}, \quad z_0 = \mathbf{c}_B^T A_B^{-1} \mathbf{b}, \quad \text{and} \quad \mathbf{r} = \mathbf{c}_N - (\mathbf{c}_B^T A_B^{-1} A_N)^T.$$

It is neither necessary nor very useful to remember these formulas; they are easily rederived if needed. The proof is not very exciting and we write it more concisely than other parts of the text and we leave some details to a diligent reader. We will proceed similarly with subsequent proofs of a similar kind.

Proof. First let us see how these formulas can be discovered: We rewrite the system $Ax = b$ to $A_B \mathbf{x}_B = \mathbf{b} - A_N \mathbf{x}_N$, and we multiply it by the inverse matrix A_B^{-1} from the left (these transformations preserve the solution set), which leads to

$$\mathbf{x}_B = A_B^{-1} \mathbf{b} - A_B^{-1} A_N \mathbf{x}_N.$$

We substitute the right-hand side for $\mathbf{x}_B$ into the equation $z = \mathbf{c}^T \mathbf{x} = \mathbf{c}_B^T \mathbf{x}_B + \mathbf{c}_N^T \mathbf{x}_N$, and we obtain

$$\begin{aligned} z &= \mathbf{c}_B^T (A_B^{-1} \mathbf{b} - A_B^{-1} A_N \mathbf{x}_N) + \mathbf{c}_N^T \mathbf{x}_N \\ &= \mathbf{c}_B^T A_B^{-1} \mathbf{b} + (\mathbf{c}_N^T - \mathbf{c}_B^T A_B^{-1} A_N) \mathbf{x}_N. \end{aligned}$$

Thus the formulas in the lemma do yield a simplex tableau, and it remains to verify the uniqueness.

Let $\mathbf{p}, Q, \mathbf{r}, z_0$ determine a simplex tableau for a feasible basis B, and let $\mathbf{p}', Q', \mathbf{r}', z_0'$ do as well. Since each choice of $\mathbf{x}_N$ determines $\mathbf{x}_B$ uniquely, the equality $\mathbf{p} + Q\mathbf{x}_N = \mathbf{p}' + Q'\mathbf{x}_N$ has to hold for all $\mathbf{x}_N \in \mathbb{R}^{n-m}$. The choice $\mathbf{x}_N = \mathbf{0}$ gives $\mathbf{p} = \mathbf{p}'$, and if we substitute the unit vectors $\mathbf{e}_j$ of the standard basis for $\mathbf{x}_N$ one by one, we also get $Q = Q'$. The equalities $z_0 = z_0'$ and $\mathbf{r} = \mathbf{r}'$ are proved similarly.   □

5.6 The Simplex Method in General

Optimality. Exactly as in the concrete example in Section 5.1, we have the following criterion of optimality of a simplex tableau:

If $\mathcal{T}(B)$ is a simplex tableau such that the coefficients of the nonbasic variables are nonpositive in the last row, i.e., if

$$\mathbf{r} \le \mathbf{0},$$

then the corresponding basic feasible solution is *optimal*.

Indeed, the basic feasible solution corresponding to such a tableau has the objective function equal to z_0, while for any other feasible solution $\tilde{\mathbf{x}}$ we have $\tilde{\mathbf{x}}_N \ge 0$ and $\mathbf{c}^T \tilde{\mathbf{x}} = z_0 + \mathbf{r}^T \tilde{\mathbf{x}}_N \le z_0$.

A pivot step: who enters and who leaves. In each step of the simplex method we go from an "old" basis B and simplex tableau $\mathcal{T}(B)$ to a "new" basis B' and the corresponding simplex tableau $\mathcal{T}(B')$. A nonbasic variable x_v enters the basis and a basic variable x_u leaves the basis,[2] and hence $B' = (B \setminus \{u\}) \cup \{v\}$.

We always select the entering variable x_v first.

A nonbasic variable may enter the basis if and only if its coefficient in the last row of the simplex tableau is *positive*.

Only incrementing such nonbasic variables increases the value of the objective function.

Usually there are several positive coefficients in the last row, and hence several possible choices of the entering variable. For the time being the reader may think of this choice as arbitrary. We will discuss ways of selecting one of these possibilities in Section 5.7.

Once we decide that the entering variable is some x_v, it remains to pick the leaving variable.

The leaving variable x_u has to be such that its nonnegativity, together with the corresponding equation in the simplex tableau having x_u on the left-hand side, limits the increment of the entering variable x_v most strictly.

Expressed by a formula, this condition might look complicated because of some double indices, but the idea is simple and we have already seen it in examples. Let us write $B = \{k_1, k_2, \ldots, k_m\}$, $k_1 < k_2 < \cdots < k_m$, and $N = \{\ell_1, \ell_2, \ldots, \ell_{n-m}\}$, $\ell_1 < \ell_2 < \cdots < \ell_{n-m}$. Then the ith equation of the simplex tableau has the form

$$x_{k_i} = p_i + \sum_{j=1}^{n-m} q_{ij} x_{\ell_j}.$$

---

[2] The letters u and v do not denote vectors here (the alphabet is not that long, after all).

We now want to write the index v of the chosen entering variable as $v = \ell_\beta$. In more detail, we define $\beta \in \{1, 2, \ldots, n-m\}$ as the index for which $v = \ell_\beta$. Similarly, the index u of the leaving variable (which hasn't been selected yet) will be written in the form $u = k_\alpha$.

Since all nonbasic variables x_{ℓ_j}, $j \neq \beta$, should remain zero, the nonnegativity condition $x_{k_i} \geq 0$ limits the possible values of the entering variable x_{ℓ_β} by the inequality $-q_{i\beta} x_{\ell_\beta} \leq p_i$. If $q_{i\beta} \geq 0$, then this inequality doesn't restrict the increase of x_{ℓ_β} in any way, while for $q_{i\beta} < 0$ it yields the restriction $x_{\ell_\beta} \leq -p_i / q_{i\beta}$.

The leaving variable x_{k_α} is thus always such that

$$q_{\alpha\beta} < 0 \quad \text{and} \quad -\frac{p_\alpha}{q_{\alpha\beta}} = \min\left\{ -\frac{p_i}{q_{i\beta}} : q_{i\beta} < 0,\ i = 1, 2, \ldots, m \right\}. \quad (5.3)$$

That is, in the simplex tableau we consider only the rows in which the coefficient of x_v is negative. In such rows we divide by this coefficient the component of the vector $\mathbf{p}$, we change sign, and we seek a minimum among these ratios. If there is no row with a negative coefficient of x_v, i.e., the minimum of the right-hand side of equation (5.3) is over an empty set, then the linear program is unbounded and the computation finishes.

For a proof that the simplex method really goes through a sequence of feasible bases we need the following lemma.

5.6.1 Lemma. *If B is a feasible basis and $\mathcal{T}(B)$ is the corresponding simplex tableau, and if the entering variable x_v and the leaving variable x_u have been selected according to the criteria described above (and otherwise arbitrarily), then $B' = (B \setminus \{u\}) \cup \{v\}$ is again a feasible basis.*

If no x_u satisfies the criterion for a leaving variable, then the linear program is unbounded. For all $t \geq 0$ we obtain a feasible solution by substituting t for x_v and 0 for all other nonbasic variables, and the value of the objective function for these feasible solutions tends to infinity as $t \to \infty$.

The proof is one of those not essential for a basic understanding of the material.

Proof (sketch). We first need to verify that the matrix $A_{B'}$ is nonsingular. This holds exactly if $A_B^{-1} A_{B'}$ is nonsingular, since we assume nonsingularity of A_B. The matrix $A_{B'}$ agrees with A_B in $m - 1$ columns corresponding to the basic variable indices $B \setminus \{u\}$. For the basic variable with index $k_i, i \neq \alpha$, we get the unit vector $\mathbf{e}_i$, in the corresponding column of $A_B^{-1} A_{B'}$.

The negative of the remaining column of the matrix $A_B^{-1} A_{B'}$ occurs in the simplex tableau $\mathcal{T}(B)$ as the column of the entering variable x_v, since $Q = -A_B^{-1} A_N$ by Lemma 5.5.1. There is a nonzero number $q_{\alpha\beta}$ in row α corresponding to the leaving variable x_u, since we have selected

x_u that way, and the other columns of $A_B^{-1} A_{B'}$ have 0 in that row. Hence the matrix is nonsingular as claimed.

Next, we need to check feasibility of the basis B'. Here we use the fact that the new basic feasible solution, that for B', can be written in terms of the old one, and the nonnegativity of its basic variables are exactly those conditions that are used for choosing the leaving variable.

In practically the same way one can show the part of the lemma dealing with unbounded linear programs. We omit further details. $\square$

A geometric view. As we saw in Section 4.4, basic feasible solutions are vertices of the polyhedron of feasible solutions. It is not hard to verify that a pivot step of the simplex method corresponds to a move from one vertex to another along an edge of the polyhedron (where an edge is a 1-dimensional face, i.e., a segment connecting the considered vertices; see Section 4.4).

Degenerate pivot steps are an exception, where we stay at the same vertex and only the feasible basis changes. A vertex of an n-dimensional convex polyhedron is generally determined by n of the bounding hyperplanes (think of a 3-dimensional cube, say). Degeneracy can occur only if we have more than n of the bounding hyperplanes meeting at a vertex (this happens for the 3-dimensional regular octahedron, for example).

Organization of the computations. Whenever we find a new feasible basis as above, we could compute the new simplex tableau according to the formulas from Lemma 5.5.1. But this is never done since it is inefficient.

For *hand calculation* the new simplex tableau is computed from the old one. We have already illustrated one possible approach in the examples. We take the equation of the old tableau with the leaving variable x_u on the left, and in this equation we carry the entering variable x_v over to the left and x_u to the right. The modified equation becomes the equation for x_v in the new tableau. The right-hand side is then substituted for x_v into all of the other equations, including the one for z in the last row. This finishes the construction of the new tableau.

In computer implementations of the simplex method, the simplex tableau is typically not computed in full. Rather, only the basic components of the basic feasible solution, i.e., the vector $\mathbf{p} = A_B^{-1}\mathbf{b}$, and the matrix A_B^{-1} are maintained. The latter allows for a fast computation of other entries of the simplex tableau when they are needed. (Let us note that for the optimality test and for selecting the entering variable we need only the last row, and for selecting the leaving variable we need only $\mathbf{p}$ and the column of the entering variable.) With respect to efficiency and numerical accuracy, the explicit inverse A_B^{-1} is not the best choice, and in practice, it is often represented by an (approximate) LU-factorization of the matrix A_B, or by other devices that can easily be updated during a pivot step of the simplex method. Since an efficient implementation of the simplex method is not among our main concerns, we will not describe how these things are actually done.

This computational approach is called the *revised simplex method*. For m considerably smaller than n it is usually much more efficient than maintaining all of the simplex tableau. In particular, $O(m^2)$ arithmetic operations per pivot step are sufficient for maintaining an LU-factorization of A_B, as opposed to about mn operations required for maintaining the simplex tableau.

Computing an initial feasible basis. If the given linear program has no "obvious" feasible basis, we look for an initial feasible basis by the procedure indicated in Section 5.4. For a linear program in the usual equational form

$$\text{maximize } \mathbf{c}^T\mathbf{x} \text{ subject to } A\mathbf{x} = \mathbf{b} \text{ and } \mathbf{x} \geq \mathbf{0}$$

we first arrange for $\mathbf{b} \geq \mathbf{0}$: We multiply the equations with $b_i < 0$ by -1. Then we introduce m new variables x_{n+1} through x_{n+m}, and we solve the auxiliary linear program

$$
\begin{aligned}
\text{maximize} \quad & -(x_{n+1} + x_{n+2} + \cdots + x_{n+m}) \\
\text{subject to} \quad & \bar{A}\bar{\mathbf{x}} = \mathbf{b} \\
& \bar{\mathbf{x}} \geq \mathbf{0},
\end{aligned}
$$

where $\bar{\mathbf{x}} = (x_1, \ldots, x_{n+m})$ is the vector of all variables including the new ones, and $\bar{A} = (A \mid I_m)$ is obtained from A by appending the $m \times m$ identity matrix to the right. The original linear program is feasible if and only if every optimal solution of the auxiliary linear program satisfies $x_{n+1} = x_{n+2} = \cdots = x_{n+m} = 0$. Indeed, it is clear that an optimal solution of the auxiliary linear program with $x_{n+1} = x_{n+2} = \cdots = x_{n+m} = 0$ yields a feasible solution of the original linear program. Conversely, any feasible solution of the original linear program provides a feasible solution of the auxiliary linear program that has the objective function equal to 0 and is thus optimal.

The auxiliary linear program can be solved by the simplex method directly, since the new variables x_{n+1} through x_{n+m} constitute an initial feasible basis. In this way we obtain some optimal solution. If it doesn't satisfy $x_{n+1} = x_{n+2} = \cdots = x_{n+m} = 0$, we are done—the original linear program is infeasible.

Let us assume that the optimal solution of the auxiliary linear program has $x_{n+1} = x_{n+2} = \cdots = x_{n+m} = 0$. The simplex method always returns a basic feasible solution. If none of the new variables x_{n+1} through x_{n+m} are in the basis for the returned optimal solution, then such a basis is then a feasible basis for the original linear program, too, and it allows us to start the simplex method.

In some degenerate cases it may happen that the basis returned by the simplex method for the auxiliary linear program contains some of the variables $x_{n+1}, \ldots, x_{n+m}$, and such a basis cannot directly be used for the original linear program. But this is a cosmetic problem only: From the returned optimal solution one can get a feasible basis for the original linear program by simple linear algebra. Namely, the optimal solution has at most m nonzero components, and their columns in the matrix A are linearly independent. If these columns are fewer than m, we can add more linearly independent columns and thus get a basis; see the proof of Lemma 4.2.1.

5.7 Pivot Rules

A **pivot rule** is a rule for selecting the entering variable if there are several possibilities, which is usually the case. Sometimes there may also be more than one possibility for choosing the leaving variable, and some pivot rules specify this choice as well, but this part is typically not so important.

The number of pivot steps needed for solving a linear program depends substantially on the pivot rule. (See the example in Section 5.1.) The problem is, of course, that we do not know in advance which choices will be good in the long run.

Here we list some of the common pivot rules. By an "improving variable" we mean any nonbasic variable with a positive coefficient in the z-row of the simplex tableau, in other words, a candidate for the entering variable.

LARGEST COEFFICIENT. Choose an improving variable with the largest coefficient in the row of the objective function z. This is the original rule, suggested by Dantzig, that maximizes the improvement of z *per unit increase* of the entering variable.

LARGEST INCREASE. Choose an improving variable that leads to the largest *absolute* improvement in z. This rule is computationally more expensive than the LARGEST COEFFICIENT rule, but it locally maximizes the progress.

STEEPEST EDGE. Choose an improving variable whose entering into the basis moves the current basic feasible solution in a direction closest to the direction of the vector $\mathbf{c}$. Written by a formula, the ratio

$$\frac{\mathbf{c}^T (\mathbf{x}_{\text{new}} - \mathbf{x}_{\text{old}})}{\|\mathbf{x}_{\text{new}} - \mathbf{x}_{\text{old}}\|}$$

should be maximized, where $\mathbf{x}_{\text{old}}$ is the basic feasible solution for the current simplex tableau and $\mathbf{x}_{\text{new}}$ is the basic feasible solution for the tableau that would be obtained by entering the considered improving variable into the basis. (We recall that $\|\mathbf{v}\| = (v_1^2 + v_2^2 + \cdots + v_n^2)^{1/2} = \sqrt{\mathbf{v}^T \mathbf{v}}$ denotes the Euclidean length of the vector $\mathbf{v}$, and the expression $\mathbf{u}^T \mathbf{v}/(\|\mathbf{u}\| \cdot \|\mathbf{v}\|)$ is the cosine of the angle of the vectors $\mathbf{u}$ and $\mathbf{v}$.)

The STEEPEST EDGE rule is a champion among pivot rules in practice. According to extensive computational studies it is usually faster than all other pivot rules described here and many others. An efficient approximate implementation of this rule is discussed in the glossary under the heading "Devex."

BLAND'S RULE. Choose the improving variable with the smallest index, and if there are several possibilities for the leaving variable, also take the one with the smallest index. BLAND'S RULE is theoretically very significant since it prevents cycling, as we will discuss in Section 5.8.

RANDOM EDGE. Select the entering variable uniformly at random among all improving variables. This is the simplest example of a *randomized pivot rule*, where the choice of the entering variable uses random numbers in some way. Randomized rules are also very important theoretically, since they lead to the current best provable bounds for the number of pivot steps of the simplex method.

5.8 The Struggle Against Cycling

As we have already mentioned, it may happen that for some linear programs the simplex method cycles (and theoretically this is the only possibility of how it may fail). Such a situation is encountered very rarely in practice, if at all, and thus many implementations simply ignore the possibility of cycling.

There are several ways that provably avoid cycling. One of them is the already mentioned BLAND'S RULE: We prove below that the simplex method never cycles if Bland's rule is applied consistently. Unfortunately, regarding efficiency, Bland's rule is one of the slowest pivot rules and it is almost never used in practice.

Another possibility can be found in the literature under the heading *lexicographic rule*, and here we only sketch it.

Cycling can occur only for degenerate linear programs. Degeneracy may lead to ties in the choice of the leaving variable. The lexicographic method breaks these ties as follows. Suppose that we have a set S of row indices such that for all $\alpha \in S$,

$$q_{\alpha\beta} < 0 \text{ and } -\frac{p_\alpha}{q_{\alpha\beta}} = \min\left\{-\frac{p_i}{q_{i\beta}} : q_{i\beta} < 0, \, i = 1, 2, \ldots, m\right\}.$$

In other words, all indices in S are candidates for the leaving variable. We then choose the index $\alpha \in S$ for which the vector

$$\left(\frac{q_{\alpha 1}}{q_{\alpha\beta}}, \ldots, \frac{q_{\alpha(n-m)}}{q_{\alpha\beta}}\right)$$

is the smallest in the lexicographic ordering. (We recall that a vector $\mathbf{x} \in \mathbb{R}^k$ is **lexicographically smaller** than a vector $\mathbf{y} \in \mathbb{R}^k$ if $x_1 < y_1$, or if $x_1 = y_1$ and $x_2 < y_2$, etc., in general, if there is an index $j \leq k$ such that $x_1 = y_1, \ldots, x_{j-1} = y_{j-1}$ and $x_j < y_j$.) Since the matrix A has rank m, it can be checked that any two of those vectors indeed differ at some index, and so we can resolve ties between any set S of rows. The chosen row index determines the leaving variable.

It can be shown that under the lexicographic rule, cycling is impossible. In very degenerate cases the lexicographic rule can be quite costly, since it may have to compute many components of the aforementioned vectors before the ties can eventually be broken.

Geometrically, the lexicographic rule has the following interpretation. For linear programs in equational form, degeneracy means that the set F of solutions of the system $A\mathbf{x} = \mathbf{b}$ contains a point with more than $n - m$ zero components, and thus it is not in general position with respect to coordinate axes. The lexicographic rule has essentially the same effect as a well-chosen perturbation of the set F, achieved by changing the vector $\mathbf{b}$ a little. This brings F into "general position" and therefore resolves all ties, while the optimal solution changes only by very little. The lexicographic rule simulates the effects of a suitable "infinitesimal" perturbation.

Now we return to Bland's rule.

5.8.1 Theorem. *The simplex method with Bland's pivot rule (the entering variable is the one with the smallest index among the eligible variables, and similarly for the leaving variable) is always finite; i.e., cycling is impossible.*

This is a basic result in the theory of linear programming (the duality theorem is an easy consequence, for example). Unfortunately, the proof is somewhat demanding. Its plot is simple, though: Assuming that there is a cycle, we get a contradiction in the form of an auxiliary linear program that has an optimal solution and is unbounded at the same time.

Proof. We assume that there is a cycle, and we let the set F consist of the indices of all variables that enter (and therefore also leave) the basis at least once during the cycle. We call these the *fickle* variables. First we verify a general claim about cycling of the simplex method, valid for any pivot rule.

Claim. All bases encountered in the cycle yield the same basic feasible solution, and all the fickle variables are 0 in it.

Proof of the claim. Since the objective function never decreases, it has to stay constant along the cycle.

Let B be a feasible basis encountered along the cycle, let $N = \{1, 2, \ldots, n\} \setminus B$ as usual, and let $B' = (B \setminus \{u\}) \cup \{v\}$ be the next basis. The only one among the *nonbasic* variables that may possibly change value in the pivot step from B to B' is the entering variable x_v; all others remain nonbasic and thus 0. By the rule for selecting the entering variable, the coefficient of x_v in the z-row of the tableau $\mathcal{T}(B)$ (i.e., in the vector $\mathbf{r}$) is strictly positive. Since the objective function is given by $z = z_0 + \mathbf{r}^T \mathbf{x}_N$, we see that if x_v became strictly positive, the objective function would increase. Hence the basic feasible solutions corresponding to B and B', respectively, agree in all components in N. Since these components determine the remaining ones uniquely (Proposition 4.2.2), the basic feasible solution does not change at all.

Finally, since every fickle variable is nonbasic at least once during the cycle, it has to be 0 all the time. The claim is proved.

The first trick in the proof of Theorem 5.8.1 is to consider the *largest* index v in the set F. Let B be a basis in the cycle just before x_v enters, and B' another basis just before x_v leaves (and x_u enters, say). Let $\mathbf{p}, Q, \mathbf{r}, z_0$ be the parameters of the simplex tableau $\mathcal{T}(B)$, and let $\mathbf{p}', Q', \mathbf{r}', z_0'$ be the parameters of $\mathcal{T}(B')$. (We remark that neither B nor B' has to be determined uniquely.)

Next, we use Bland's rule to infer some properties of the tableaus $\mathcal{T}(B)$ and $\mathcal{T}(B')$. First we focus on the situation at B. As in Section 5.6, we write B and $N = \{1, 2, \ldots, n\} \setminus B$ as ordered sets: $B = \{k_1, k_2, \ldots, k_m\}$, $k_1 < k_2 < \cdots < k_m$, and $N = \{\ell_1, \ell_2, \ldots, \ell_{n-m}\}$, $\ell_1 < \ell_2 < \cdots < \ell_{n-m}$. Since we have chosen v as the largest index in F, and Bland's rule requires v to be the smallest index of a candidate for entering the basis, no other fickle variable is a candidate at this point. Thus all fickle variables except for x_v have nonpositive coefficients in the z-row of $\mathcal{T}(B)$. Expressed formally, if β is the index such that $v = \ell_\beta$, we have

$$r_\beta > 0 \text{ and } r_j \leq 0 \text{ for all } j \text{ such that } \ell_j \in F \setminus \{v\}. \tag{5.4}$$

Second, we consider the tableau $\mathcal{T}(B')$. We write $B' = \{k_1', k_2', \ldots, k_m'\}$, $N' = \{1, 2, \ldots, n\} \setminus B' = \{\ell_1', \ell_2', \ldots, \ell_{n-m}'\}$, we let α' be the index of the leaving variable x_v in B', i.e., the one with $k_{\alpha'}' = v$, and we let β' be the index of the entering variable x_u in N', i.e., the one with $\ell_{\beta'}' = u$. By the same logic as above, x_v is the *only* candidate for leaving the basis among all the basic fickle variables in $\mathcal{T}(B')$. Recalling the criterion (5.3) for leaving the basis, we get that $i = \alpha'$ is the only i with $k_i' \in F$ and $q_{i\beta'}' < 0$ that minimizes the ratio $-p_i'/q_{i\beta'}'$. Since $\mathbf{p}'$ specifies the values of the basic variables and all fickle variables remain 0 during the cycle, we have $p_i' = 0$ for all i with $k_i' \in F$. Consequently,

$$q_{\alpha'\beta'}' < 0 \text{ and } q_{i\beta'}' \geq 0 \text{ for all } i \text{ such that } k_i' \in F \setminus \{v\}. \tag{5.5}$$

The idea is now to construct an auxiliary linear program for which (5.4) proves that it has an optimal solution, while (5.5) shows that it is unbounded. This is a clear contradiction, which rules out the initial assumption, the existence of a cycle.

The auxiliary linear program is the following:

$$\text{Maximize} \quad \mathbf{c}^T \mathbf{x}$$
$$\text{subject to} \quad A\mathbf{x} = \mathbf{b}$$
$$\mathbf{x}_{F \setminus \{v\}} \geq \mathbf{0}$$
$$x_v \leq 0$$
$$\mathbf{x}_{N \setminus F} = \mathbf{0}.$$

We want stress that here the variables $\mathbf{x}_{B \setminus F}$ may assume *any* signs.

Optimality of the auxiliary linear program. We let $\tilde{\mathbf{x}}$ be the basic feasible solution of our original linear program associated with the basis B. Since $\tilde{\mathbf{x}}_N = \mathbf{0}$ and $\tilde{\mathbf{x}}_F = \mathbf{0}$ (by the claim), $\tilde{\mathbf{x}}$ is feasible for the auxiliary program. Moreover, for every $\mathbf{x}$ satisfying $A\mathbf{x} = \mathbf{b}$ the value of the objective function can be expressed as

$$\mathbf{c}^T \mathbf{x} = z = z_0 + \mathbf{r}^T \mathbf{x}_N.$$

For all feasible solutions $\mathbf{x}$ of the auxiliary linear program, we have

$$x_{\ell_j} \begin{cases} \geq 0 & \text{if } \ell_j \in F \setminus \{v\} \\ \leq 0 & \text{if } \ell_j = \ell_\beta = v, \end{cases}$$

and so (5.4) implies

$$r_j x_{\ell_j} \leq 0 \text{ for all } j \text{ such that } \ell_j \in F.$$

Together with $\mathbf{x}_{N \setminus F} = \mathbf{0}$, we get $\mathbf{r}^T \mathbf{x}_N \leq \mathbf{0}$, and hence $z \leq z_0$ for all feasible solutions of the auxiliary linear program. It follows that $\tilde{\mathbf{x}}$ is an optimal solution of the auxiliary linear program.

Unboundedness of the auxiliary linear program. By the claim at the beginning of the proof, $\tilde{\mathbf{x}}$ is also the basic feasible solution of our original linear program associated with the basis B'. For all solutions $\mathbf{x}$ of $A\mathbf{x} = \mathbf{b}$ we have

$$\mathbf{x}_{B'} = \mathbf{p}' + Q' \mathbf{x}_{N'}. \tag{5.6}$$

Let us now change $\tilde{\mathbf{x}}_{N'}$, by letting $\tilde{x}_u$ grow from its current value 0 to some value $t > 0$. Using (5.6), this determines a new solution $\tilde{\mathbf{x}}(t)$ of $A\mathbf{x} = \mathbf{b}$; we will show that for all $t > 0$, this solution is feasible for the auxiliary problem, but that the objective function value $\mathbf{c}^T \tilde{\mathbf{x}}(t)$ tends to infinity as $t \to \infty$. Here are the details.

We set

$$\tilde{x}_{\ell'_j}(t) := \begin{cases} 0 & \text{if } \ell'_j \in N' \setminus u \\ t & \text{if } \ell'_j = \ell'_{\beta'} = u. \end{cases}$$

With $\tilde{x}_v = 0$ and $t > 0$, (5.6) and (5.5) together show that

$$\tilde{x}_{k'_i}(t) = \tilde{x}_{k'_i} + t q'_{i\beta'} \begin{cases} \geq 0 & \text{if } k'_i \in F \setminus \{v\} \\ < 0 & \text{if } k'_i = k'_{\alpha'} = v. \end{cases}$$

In particular, $\tilde{\mathbf{x}}(t)$ is again feasible for the auxiliary linear program.

Since the variable $x_u = x_{\ell'_{\beta'}}$ was a candidate for entering the basis B', we know that $r'_{\beta'} > 0$, and hence

$$\mathbf{c}^T \tilde{\mathbf{x}}(t) = z'_0 + \mathbf{r}'^T \tilde{\mathbf{x}}_{N'}(t) = z'_0 + t r'_{\beta'} \to \infty \quad \text{for } t \to \infty.$$

This means that the auxiliary linear program is unbounded.   □

5.9 Efficiency of the Simplex Method

In practice, the simplex method performs very satisfactorily even for large linear programs. Computational experiments indicate that for linear programs in equational form with m equations it typically reaches an optimal solution in something between $2m$ and $3m$ pivot steps.

It was thus a great surprise when Klee and Minty constructed a linear program with n nonnegative variables and n inequalities for which the simplex method with Dantzig's original pivot rule (LARGEST COEFFICIENT) needs exponentially many pivot steps, namely $2^n - 1$!

The set of feasible solutions is an ingeniously deformed n-dimensional cube, called the *Klee–Minty cube*, constructed in such a way that the simplex method passes through all of its vertices. It is not hard to see that there is a deformed n-dimensional cube with an x_n-increasing path, say, through all vertices. Instead of a formal description we illustrate such a construction by pictures for dimensions 2 and 3:

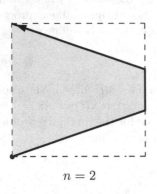

$n = 2$

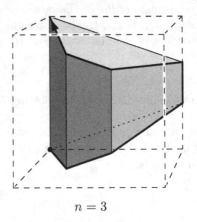

$n = 3$

The deformed cube is inscribed in an ordinary cube in order to better convey the shape. With some pivot rules, the simplex method may traverse the path marked with a thick line. The particular deformed

cube shown in the picture won't fool Dantzig's rule, for which the original example of Klee and Minty was constructed, though. A deformed
cube that does fool Dantzig's rule looks more bizarre:

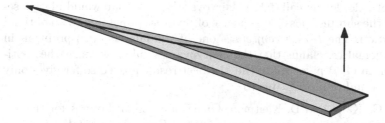

The direction of the objective function is drawn vertically. The corresponding linear program with $n = 3$ variables is simple:

$$
\begin{array}{lrcl}
\text{Maximize} & 9x_1 + 3x_2 + x_3 & & \\
\text{subject to} & x_1 & \leq & 1 \\
& 6x_1 + x_2 & \leq & 9 \\
& 18x_1 + 6x_2 + x_3 & \leq & 81 \\
& x_1, x_2, x_3 & \geq & 0.
\end{array}
$$

It is instructive to see how, after the standard conversion to equational
form, this linear program forces Dantzig's rule to go through all feasible bases.

Later on, very slow examples of a similar type were discovered for many
other pivot rules, among them all the rules mentioned above. Many people
have tried to design a pivot rule and prove that the number of pivot steps is
always bounded by some polynomial function of m and n, but nobody has
succeeded so far. The best known bound has been proved for the following
simple randomized pivot rule: Choose a random ordering of the variables at
the beginning of the computation (in other words, randomly permute the
indices of the variables in the input linear program); then use Bland's rule
for choosing the entering variable, and the lexicographic method for choosing
the leaving variable. For every linear program with at most n variables and
at most n constraints, the expected number of pivot steps is bounded by
$e^{C\sqrt{n \ln n}}$, where C is a (not too large) constant. (Here the expectation means
the arithmetic average over all possible orderings of the variables.) This bound
is considerably better than 2^n, say, but much worse than a polynomial bound.

This algorithm was found independently and almost at the same
time by Kalai and by Matoušek, Sharir, and Welzl. For a recent treatment in a somewhat broader context see

B. Gärtner and E. Welzl: Explicit and implicit enforcing—randomized optimization, in *Lectures of the Graduate Program*

Computational Discrete Mathematics, Lecture Notes in Computer Science 2122 (2001), Springer, Berlin etc., pages 26–49.

A very good bound is not known even for the cleverest possible pivot rule, let us call it the clairvoyant's rule, that would always select the shortest possible sequence of pivot steps leading to an optimal solution. The *Hirsch conjecture*, one of the famous open problems in mathematics, claims that the clairvoyant's rule always reaches optimum in $O(n)$ pivot steps. But the best result proved so far gives only the bound of $n^{1+\ln n}$, due to

G. Kalai and D. Kleitman: Quasi-polynomial bounds for the diameter of graphs of polyhedra, *Bull. Amer Math. Soc.* 26(1992), 315–316.

This is better than $e^{C\sqrt{n \ln n}}$, but still worse than any polynomial function of n, and it doesn't provide a real pivot rule since nobody knows how to simulate clairvoyant's decisions by an efficient algorithm.

Here is an approach that looks promising and has been tried more recently, although without a clear success so far. One tries to modify the given linear program in such a way that polynomiality of a suitable pivot rule for the modified linear program would be easier to prove, and of course, so that an optimal solution of the original linear program could easily be derived from an optimal solution of the modified linear program.

In spite of the Klee–Minty cube and similar artificial examples, the simplex method is being used successfully. Remarkable theoretical results indicate that these willful examples are rare indeed. For instance, it is known that if a linear program in equational form is generated in a suitable (precisely defined) way at random, then the number of pivot steps is of order at most m^2 with high probability. More recent results, in the general framework of the so-called *smoothed complexity*, claim that if we take an arbitrary linear program and then we change its coefficients by small random amounts, then the simplex method with a certain pivot rule reaches the optimum of the resulting linear program by polynomially many steps with high probability (a concrete bound on the polynomial depends on a precise specification of the "small random amounts" of change). The first theorem of this kind is due to Spielman and Teng, and for recent progress see

R. Vershynin: Beyond Hirsch conjecture: Walks on random polytopes and the smoothed complexity of the simplex method, preprint, 2006.

An exact formulation of these results requires a number of rather technical notions that we do not want to introduce here, and so we omit it.

5.10 Summary

Let us review the simplex method once again.

Algorithm SIMPLEX METHOD

1. Convert the input linear program to equational form

   $$\text{maximize } \mathbf{c}^T \mathbf{x} \text{ subject to } A\mathbf{x} = \mathbf{b} \text{ and } \mathbf{x} \geq \mathbf{0}$$

   with n variables and m equations, where A has rank m (see Section 4.1).
2. If no feasible basis is available, arrange for $\mathbf{b} \geq \mathbf{0}$, and solve the following auxiliary linear program by the simplex method:

   $$\begin{aligned} \text{Maximize} \quad & -(x_{n+1} + x_{n+2} + \cdots + x_{n+m}) \\ \text{subject to} \quad & \bar{A}\,\bar{\mathbf{x}} = \mathbf{b} \\ & \bar{\mathbf{x}} \geq \mathbf{0}, \end{aligned}$$

   where $x_{n+1}, \ldots, x_{n+m}$ are new variables, $\bar{\mathbf{x}} = (x_1, \ldots, x_{n+m})$, and $\bar{A} = (A \mid I_m)$. If the optimal value of the objective function comes out negative, the original linear program is infeasible; **stop**. Otherwise, the first n components of the optimal solution form a basic feasible solution of the original linear program.
3. For a feasible basis $B \subseteq \{1, 2, \ldots, n\}$ compute the simplex tableau $\mathcal{T}(B)$, of the form

   $$\begin{aligned} \mathbf{x}_B &= \mathbf{p} + Q\mathbf{x}_N \\ \hline z &= z_0 + \mathbf{r}^T \mathbf{x}_N \end{aligned}$$

4. If $\mathbf{r} \leq \mathbf{0}$ in the current simplex tableau, return an optimal solution ($\mathbf{p}$ specifies the basic components, while the nonbasic components are 0); **stop**.
5. Otherwise, select an *entering variable* x_v whose coefficient in the vector $\mathbf{r}$ is positive. If there are several possibilities, use some pivot rule.
6. If the column of the entering variable x_v in the simplex tableau is non-negative, the linear program is unbounded; **stop**.
7. Otherwise, select a *leaving variable* x_u. Consider all rows of the simplex tableau where the coefficient of x_v is negative, and in each such row divide the component of the vector $\mathbf{p}$ by that coefficient and change sign. The row of the leaving variable is one in which this ratio is minimal. If there are several possibilities, decide by a pivot rule, or arbitrarily if the pivot rule doesn't specify how to break ties in this case.
8. Replace the current feasible basis B by the new feasible basis $(B \setminus \{u\}) \cup \{v\}$. Update the simplex tableau so that it corresponds to this new basis. Go to Step 4.

This is all we wanted to say about the simplex method here. May your pivot steps lead you straight to the optimum and never cycle!

6. Duality of Linear Programming

6.1 The Duality Theorem

Here we formulate arguably the most important theoretical result about linear programs.

Let us consider the linear program

$$
\begin{aligned}
\text{maximize} \quad & 2x_1 + 3x_2 \\
\text{subject to} \quad & 4x_1 + 8x_2 \leq 12 \\
& 2x_1 + x_2 \leq 3 \\
& 3x_1 + 2x_2 \leq 4 \\
& x_1, x_2 \geq 0.
\end{aligned}
\tag{6.1}
$$

Without computing the optimum, we can immediately infer from the first inequality and from the nonnegativity constraints that the maximum of the objective function is not larger than 12, because for nonnegative x_1 and x_2 we have

$$2x_1 + 3x_2 \leq 4x_1 + 8x_2 \leq 12.$$

We obtain a better upper bound if we first divide the first inequality by two:

$$2x_1 + 3x_2 \leq 2x_1 + 4x_2 \leq 6.$$

An even better bound results if we add the first two inequalities together and divide by three, which leads to the inequality

$$2x_1 + 3x_2 = \frac{1}{3}(4x_1 + 8x_2 + 2x_1 + x_2) \leq \frac{1}{3}(12 + 3) = 5,$$

and hence the objective function cannot be larger than 5.

How good an upper bound can we get in this way? And what does "in this way" mean? Let us begin with the latter question: From the constraints, we are trying to derive an inequality of the form

$$d_1 x_1 + d_2 x_2 \leq h,$$

where $d_1 \geq 2$, $d_2 \geq 3$, and h is as small as possible. Then we can claim that for all $x_1, x_2 \geq 0$ we have

$$2x_1 + 3x_2 \le d_1 x_1 + d_2 x_2 \le h,$$

and therefore, h is an upper bound on the maximum of the objective function. How can we derive such inequalities? We combine the three inequalities in the linear program with some nonnegative coefficients y_1, y_2, y_3 (nonnegativity is needed so that the direction of inequality is not reversed). We obtain

$$(4y_1 + 2y_2 + 3y_3)x_1 + (8y_1 + y_2 + 2y_3)x_2 \le 12y_1 + 3y_2 + 4y_3,$$

and thus $d_1 = 4y_1 + 2y_2 + 3y_3$, $d_2 = 8y_1 + y_2 + 2y_3$, and $h = 12y_1 + 3y_2 + 4y_3$.

How do we choose the best coefficients y_1, y_2, y_3? We must ensure that $d_1 \ge 2$ and $d_2 \ge 3$, and we want h to be as small as possible under these constraints. This is again a linear program:

$$
\begin{array}{ll}
\text{Minimize} & 12y_1 + 3y_2 + 4y_3 \\
\text{subject to} & 4y_1 + 2y_2 + 3y_3 \ge 2 \\
& 8y_1 + y_2 + 2y_3 \ge 3 \\
& y_1, y_2, y_3 \ge 0.
\end{array}
$$

It is called the linear program *dual* to the linear program (6.1) we started with. The dual linear program "guards" the original linear program from above, in the sense that every feasible solution (y_1, y_2, y_3) of the dual linear program provides an upper bound on the maximum of the objective function in (6.1).

How well does it guard? Perfectly! The optimal solution of the dual linear program is $\mathbf{y} = (\frac{5}{16}, 0, \frac{1}{4})$ with objective function equal to 4.75, and this is also the optimal value of the linear program (6.1), which is attained for $\mathbf{x} = (\frac{1}{2}, \frac{5}{4})$.

The duality theorem asserts that the dual linear program *always* guards perfectly. Let us repeat the above considerations in a more general setting, for a linear program of the form

$$\text{maximize } \mathbf{c}^T \mathbf{x} \text{ subject to } A\mathbf{x} \le \mathbf{b} \text{ and } \mathbf{x} \ge \mathbf{0}, \tag{P}$$

where A is a matrix with m rows and n columns. We are trying to combine the m inequalities of the system $A\mathbf{x} \le \mathbf{b}$ with some nonnegative coefficients $y_1, y_2, \ldots, y_m$ so that

- the resulting inequality has the jth coefficient at least c_j, $j = 1, 2, \ldots, n$, and
- the right-hand side is as small as possible.

This leads to the **dual linear program**

$$\text{minimize } \mathbf{b}^T \mathbf{y} \text{ subject to } A^T \mathbf{y} \ge \mathbf{c} \text{ and } \mathbf{y} \ge \mathbf{0}; \tag{D}$$

whoever doesn't believe this may write it in components. In this context the linear program (P) is referred to as the **primal linear program**.

From the way we have produced the dual linear program (D), we obtain the following result:

6.1.1 Proposition. *For each feasible solution* $\mathbf{y}$ *of the dual linear program* (D) *the value* $\mathbf{b}^T\mathbf{y}$ *provides an upper bound on the maximum of the objective function of the linear program* (P). *In other words, for each feasible solution* $\mathbf{x}$ *of* (P) *and each feasible solution* $\mathbf{y}$ *of* (D) *we have*

$$\mathbf{c}^T\mathbf{x} \leq \mathbf{b}^T\mathbf{y}.$$

In particular, if (P) *is unbounded,* (D) *has to be infeasible, and if* (D) *is unbounded (from below!), then* (P) *is infeasible.*

This proposition is usually called the *weak duality theorem*, weak because it expresses only the guarding of the primal linear program (P) by the dual linear program (D), but it doesn't say that the guarding is perfect. The latter is expressed only by the duality theorem (sometimes also called the *strong duality theorem*).

Duality theorem of linear programming

For the linear programs

$$\text{maximize } \mathbf{c}^T\mathbf{x} \text{ subject to } A\mathbf{x} \leq \mathbf{b} \text{ and } \mathbf{x} \geq \mathbf{0} \qquad \text{(P)}$$

and

$$\text{minimize } \mathbf{b}^T\mathbf{y} \text{ subject to } A^T\mathbf{y} \geq \mathbf{c} \text{ and } \mathbf{y} \geq \mathbf{0} \qquad \text{(D)}$$

exactly one of the following possibilities occurs:

1. Neither (P) nor (D) has a feasible solution.
2. (P) is unbounded and (D) has no feasible solution.
3. (P) has no feasible solution and (D) is unbounded.
4. Both (P) and (D) have a feasible solution. Then both have an optimal solution, and if $\mathbf{x}^*$ is an optimal solution of (P) and $\mathbf{y}^*$ is an optimal solution of (D), then

$$\mathbf{c}^T\mathbf{x}^* = \mathbf{b}^T\mathbf{y}^*.$$

That is, the *maximum of* (P) *equals the minimum of* (D).

The duality theorem might look complicated at first encounter. For understanding it better it may be useful to consider a simpler version, called the Farkas lemma and discussed in Section 6.4. This simpler statement has several intuitive interpretations, and it contains the essence of the duality theorem.

Proving the duality theorem, which we will undertake in Sections 6.3 and 6.4, does take some work, unlike a proof of the weak duality theorem, which is quite easy.

The heart of the duality theorem is the equality $\mathbf{c}^T\mathbf{x}^* = \mathbf{b}^T\mathbf{y}^*$ in the fourth possibility, i.e., for both (P) and (D) feasible.

Since a linear program can be either feasible and bounded, or feasible and unbounded, or infeasible, there are 3 possibilities for (P) and 3 possibilities for (D), which at first sight gives 9 possible combinations for (P) and (D). The three cases "(P) unbounded and (D) feasible bounded," "(P) unbounded and (D) unbounded," and "(P) feasible bounded and (D) unbounded" are ruled out by the weak duality theorem. In the proof of the duality theorem, we will rule out the cases "(P) infeasible and (D) feasible bounded," as well as "(P) feasible bounded and (D) infeasible." This leaves us with the four cases listed in the duality theorem. All of them can indeed occur.

Once again: feasibility versus optimality. In Chapter 1 we remarked that finding a feasible solution of a linear program is in general computationally as difficult as finding an optimal solution. There we briefly substantiated this claim using binary search. The duality theorem provides a considerably more elegant argument: The linear program (P) has an optimal solution if and only if the following linear program, obtained by combining the constraints of (P), the constraints of (D), and an inequality between the objective functions, has a feasible solution:

$$
\begin{array}{ll}
\text{Maximize} & \mathbf{c}^T\mathbf{x} \\
\text{subject to} & A\mathbf{x} \le \mathbf{b}, \\
& A^T\mathbf{y} \ge \mathbf{c}, \\
& \mathbf{c}^T\mathbf{x} \ge \mathbf{b}^T\mathbf{y}, \\
& \mathbf{x} \ge \mathbf{0}, \mathbf{y} \ge \mathbf{0}.
\end{array}
$$

(the objective function is immaterial here, and the variables are $x_1, \ldots, x_n$ and $y_1, \ldots, y_m$). Moreover, for each feasible solution $(\tilde{\mathbf{x}}, \tilde{\mathbf{y}})$ of the last linear program, $\tilde{\mathbf{x}}$ is an optimal solution of the linear program (P). All of this is a simple consequence of the duality theorem.

6.2 Dualization for Everyone

The duality theorem is valid for each linear program, not only for one of the form (P); we have only to construct the dual linear program properly. To this end, we can convert the given linear program to the form (P) using the tricks from Sections 1.1 and 4.1, and then the dual linear program has the form (D). The result can often be simplified; for example, the difference of two nonnegative variables can be replaced by a single unbounded variable (one that may attain all real values).

Simpler than doing this again and again is to adhere to the recipe below (whose validity can be proved by the just mentioned procedure). Let us assume that the primal linear program has variables $x_1, x_2, \ldots, x_n$, among which some may be nonnegative, some nonpositive, and some unbounded. Let the constraints be $C_1, C_2, \ldots, C_m$, where C_i has the form

$$a_{i1}x_1 + a_{i2}x_2 + \cdots + a_{in}x_n \begin{Bmatrix} \leq \\ \geq \\ = \end{Bmatrix} b_i.$$

(The nonnegativity or nonpositivity constraints for the variables are not counted among the C_i.) The objective function $\mathbf{c}^T\mathbf{x}$ should be *maximized*.

Then the dual linear program has variables $y_1, y_2, \ldots, y_m$, where y_i corresponds to the constraint C_i and satisfies

$$\begin{Bmatrix} y_i \geq 0 \\ y_i \leq 0 \\ y_i \in \mathbb{R} \end{Bmatrix} \text{ if we have } \begin{Bmatrix} \leq \\ \geq \\ = \end{Bmatrix} \text{ in } C_i.$$

The constraints of the dual linear program are $Q_1, Q_2, \ldots, Q_n$, where Q_j corresponds to the variable x_j and reads

$$a_{1j}y_1 + a_{2j}y_2 + \cdots + a_{mj}y_m \begin{Bmatrix} \geq \\ \leq \\ = \end{Bmatrix} c_j \text{ if } x_j \text{ satisfies } \begin{Bmatrix} x_j \geq 0 \\ x_j \leq 0 \\ x_j \in \mathbb{R} \end{Bmatrix}.$$

The objective function is $\mathbf{b}^T\mathbf{y}$, and it is to be *minimized*.

Note that in the first part of the recipe (from primal constraints to dual variables) the direction of inequalities is reversed, while in the second part (from primal variables to dual constraints) the direction is preserved.

Dualization Recipe

	Primal linear program	Dual linear program
Variables	$x_1, x_2, \ldots, x_n$	$y_1, y_2, \ldots, y_m$
Matrix	A	A^T
Right-hand side	$\mathbf{b}$	$\mathbf{c}$
Objective function	$\max \mathbf{c}^T\mathbf{x}$	$\min \mathbf{b}^T\mathbf{y}$
Constraints	ith constraint has $\leq$ $\geq$ $=$	$y_i \geq 0$ $y_i \leq 0$ $y_i \in \mathbb{R}$
	$x_j \geq 0$ $x_j \leq 0$ $x_j \in \mathbb{R}$	jth constraint has $\geq$ $\leq$ $=$

If we want to dualize a *minimization* linear program, we can first transform it to a maximization linear program by changing the sign of the objective function, and then follow the recipe.

In this way one can also find out that the rules work symmetrically "there" and "back." By this we mean that if we start with some linear program, construct the dual linear program, and then again the dual linear program, we get back to the original (primal) linear program; two consecutive dualizations cancel out. In particular, the linear programs (P) and (D) in the duality theorem are *dual to each other*.

A physical interpretation of duality. Let us consider a linear program

$$\text{maximize } \mathbf{c}^T \mathbf{x} \text{ subject to } A\mathbf{x} \le \mathbf{b}.$$

According to the dualization recipe the dual linear program is

$$\text{minimize } \mathbf{b}^T \mathbf{y} \text{ subject to } A^T \mathbf{y} = \mathbf{c} \text{ and } \mathbf{y} \ge \mathbf{0}.$$

Let us assume that the primal linear program is feasible and bounded, and let $n = 3$. We regard $\mathbf{x}$ as a point in three-dimensional space, and we interpret $\mathbf{c}$ as the gravitation vector; it thus points downward.

Each of the inequalities of the system $A\mathbf{x} \le \mathbf{b}$ determines a half-space. The intersection of these half-spaces is a nonempty convex polyhedron bounded from below. Each of its two-dimensional faces is given by one of the equations $\mathbf{a}_i^T \mathbf{x} = b_i$, where the vectors $\mathbf{a}_1, \mathbf{a}_2, \ldots, \mathbf{a}_m$ are the rows of the matrix A, but interpreted as column vectors. Let us denote the face given by $\mathbf{a}_i^T \mathbf{x} = b_i$ by S_i (not every inequality of the system $A\mathbf{x} \le \mathbf{b}$ has to correspond to a face, and so S_i is not necessarily defined for every i).

Let us imagine that the boundary of the polyhedron is made of cardboard and that we drop a tiny steel ball somewhere inside the polyhedron. The ball falls and rolls down to the lowest vertex (or possibly it stays on a horizontal edge or face). Let us denote the resulting position of the ball by $\mathbf{x}^*$; thus, $\mathbf{x}^*$ is an optimal solution of the linear program. In this stable position the ball touches several two-dimensional faces, typically 3. Let D be the set of i such that the ball touches the face S_i. For $i \in D$ we thus have

$$\mathbf{a}_i^T \mathbf{x}^* = b_i. \tag{6.2}$$

Gravity exerts a force $\mathbf{F}$ on the ball that is proportional to the vector $\mathbf{c}$. This force is decomposed into forces of pressure on the faces touched by the ball. The force $\mathbf{F}_i$ by which the ball acts on face S_i is orthogonal to S_i and it is directed outward from the polyhedron (if we neglect friction); see the schematic two-dimensional picture below:

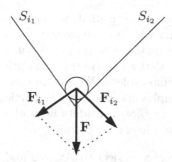

The forces acting on the ball are in equilibrium, and thus $\mathbf{F} = \sum_{i \in D} \mathbf{F}_i$. The outward normal of the face S_i is $\mathbf{a}_i$; hence $\mathbf{F}_i$ is proportional to $\mathbf{a}_i$, and for some nonnegative numbers y_i^* we have

$$\sum_{i \in D} y_i^* \mathbf{a}_i = \mathbf{c}.$$

If we set $y_i^* = 0$ for $i \notin D$, we can write $\sum_{i=1}^{m} y_i^* \mathbf{a}_i = \mathbf{c}$, or $A^T \mathbf{y}^* = \mathbf{c}$ in matrix form. Therefore, $\mathbf{y}^*$ is a feasible solution of the dual linear program.

Let us consider the product $(\mathbf{y}^*)^T (A \mathbf{x}^* - \mathbf{b})$. For $i \notin D$ the ith component of $\mathbf{y}^*$ equals 0, while for $i \in D$ the ith component of $A \mathbf{x}^* - \mathbf{b}$ is 0 according to (6.2). So the product is 0, and hence $(\mathbf{y}^*)^T \mathbf{b} = (\mathbf{y}^*)^T A \mathbf{x}^* = \mathbf{c}^T \mathbf{x}^*$.

We see that $\mathbf{x}^*$ is a feasible solution of the primal linear program, $\mathbf{y}^*$ is a feasible solution of the dual linear program, and $\mathbf{c}^T \mathbf{x}^* = \mathbf{b}^T \mathbf{y}^*$. By the weak duality theorem $\mathbf{y}^*$ is an optimal solution of the dual linear program, and we have a situation exactly as in the duality theorem. We have just "physically verified" a special three-dimensional case of the duality theorem.

We remark that the dual linear program also has an economic interpretation. The dual variables are called *shadow prices* in this context. The interested reader will find this nicely explained in Chvátal's textbook cited in Chapter 9.

6.3 Proof of Duality from the Simplex Method

The duality theorem of linear programming can be quickly derived from the correctness of the simplex method. To be precise, we will prove the following:

If the primal linear program (P) *is feasible and bounded, then the dual linear program* (D) *is feasible (and bounded as well, by weak duality), with the same optimum value as the primal.*

Since the dual of the dual is the primal, we may interchange (P) and (D) in this statement. Together with our considerations about the possible cases after the statement of the duality theorem, this will prove the theorem.

The key observation is that we can extract an optimal solution of the dual linear program from the final tableau. We should recall, though, that proving the correctness of the simplex method, and in particular, the fact that one can always avoid cycling, requires considerable work.

Let us consider a primal linear program

$$\text{maximize } \mathbf{c}^T \mathbf{x} \text{ subject to } A\mathbf{x} \le \mathbf{b} \text{ and } \mathbf{x} \ge \mathbf{0}. \tag{P}$$

After a conversion to equational via slack variables $x_{n+1}, \ldots, x_{n+m}$ we arrive at the linear program

$$\text{maximize } \bar{\mathbf{c}}^T \bar{\mathbf{x}} \text{ subject to } \bar{A}\bar{\mathbf{x}} = \mathbf{b} \text{ and } \bar{\mathbf{x}} \ge \mathbf{0},$$

where $\bar{\mathbf{x}} = (x_1, \ldots, x_{n+m})$, $\bar{\mathbf{c}} = (c_1, \ldots, c_n, 0, \ldots, 0)$, and $\bar{A} = (A \mid I_m)$. If this last linear program is feasible and bounded, then according to Theorem 5.8.1, the simplex method with Bland's rule always finds some optimal solution $\bar{\mathbf{x}}^*$ with a feasible basis B. The first n components of the vector $\bar{\mathbf{x}}^*$ constitute an optimal solution $\mathbf{x}^*$ of the linear program (P). By the optimality criterion we have $\mathbf{r} \le \mathbf{0}$ in the final simplex tableau, where $\mathbf{r}$ is the vector in the z-row of the tableau as in Section 5.5. The following lemma and the weak duality theorem (Proposition 6.1.1) then easily imply the duality theorem.

6.3.1 Lemma. *In the described situation the vector* $\mathbf{y}^* = (\bar{\mathbf{c}}_B^T \bar{A}_B^{-1})^T$ *is a feasible solution of the dual linear program* (D) *and the equality* $\mathbf{c}^T \mathbf{x}^* = \mathbf{b}^T \mathbf{y}^*$ *holds.*

Proof. By Lemma 5.5.1, $\bar{\mathbf{x}}^*$ is given by $\bar{\mathbf{x}}_B^* = \bar{A}_B^{-1}\mathbf{b}$ and $\bar{\mathbf{x}}_N^* = \mathbf{0}$, and so

$$\mathbf{c}^T \mathbf{x}^* = \bar{\mathbf{c}}^T \bar{\mathbf{x}}^* = \bar{\mathbf{c}}_B^T \bar{\mathbf{x}}_B^* = \bar{\mathbf{c}}_B^T (\bar{A}_B^{-1}\mathbf{b}) = (\bar{\mathbf{c}}_B^T \bar{A}_B^{-1})\mathbf{b} = (\mathbf{y}^*)^T \mathbf{b} = \mathbf{b}^T \mathbf{y}^*.$$

The equality $\mathbf{c}^T \mathbf{x}^* = \mathbf{b}^T \mathbf{y}^*$ thus holds, and it remains to check the feasibility of $\mathbf{y}^*$, that is, $A^T \mathbf{y}^* \ge \mathbf{c}$ and $\mathbf{y}^* \ge \mathbf{0}$.

The condition $\mathbf{y}^* \ge \mathbf{0}$ can be rewritten to $I_m \mathbf{y}^* \ge \mathbf{0}$, and hence both of the feasibility conditions together are equivalent to

$$\bar{A}^T \mathbf{y}^* \ge \bar{\mathbf{c}}. \tag{6.3}$$

After substituting $\mathbf{y}^* = (\bar{\mathbf{c}}_B^T \bar{A}_B^{-1})^T$ the left-hand side becomes $\bar{A}^T (\bar{\mathbf{c}}_B^T \bar{A}_B^{-1})^T = (\bar{\mathbf{c}}_B^T \bar{A}_B^{-1} \bar{A})^T$. Let us denote this $(n+m)$-component vector by $\mathbf{w}$. For the basic components of $\mathbf{w}$ we have

$$\mathbf{w}_B = (\bar{\mathbf{c}}_B^T \bar{A}_B^{-1} \bar{A}_B)^T = (\bar{\mathbf{c}}_B^T I_m)^T = \bar{\mathbf{c}}_B,$$

and thus we even have equality in (6.3) for the basic components. For the nonbasic components we have

$$\mathbf{w}_N = (\bar{\mathbf{c}}_B^T \bar{A}_B^{-1} \bar{A}_N)^T = \bar{\mathbf{c}}_N - \mathbf{r} \geq \bar{\mathbf{c}}_N$$

since $\mathbf{r} = \bar{\mathbf{c}}^N - (\bar{\mathbf{c}}_B^T \bar{A}_B^{-1} \bar{A}_N)^T$ by Lemma 5.5.1, and $\mathbf{r} \leq \mathbf{0}$ by the optimality criterion. The lemma is proved. $\qquad\qquad\qquad\qquad\qquad\qquad\qquad$ □

6.4 Proof of Duality from the Farkas Lemma

Another approach to the duality theorem of linear programming consists in first proving a simplified version, called the *Farkas lemma*, and then substituting a skillfully composed matrix into it and thus deriving the theorem. A nice feature is that the Farkas lemma has very intuitive interpretations.

Actually, the Farkas lemma comes in several natural variants. We begin by discussing one of them, which has a very clear geometric meaning.

6.4.1 Proposition (Farkas lemma). *Let A be a real matrix with m rows and n columns, and let $\mathbf{b} \in \mathbb{R}^m$ be a vector. Then exactly one of the following two possibilities occurs:*

(F1) *There exists a vector $\mathbf{x} \in \mathbb{R}^n$ satisfying $A\mathbf{x} = \mathbf{b}$ and $\mathbf{x} \geq \mathbf{0}$.*
(F2) *There exists a vector $\mathbf{y} \in \mathbb{R}^m$ such that $\mathbf{y}^T A \geq \mathbf{0}^T$ and $\mathbf{y}^T \mathbf{b} < 0$.*

It is easily seen that both possibilities cannot occur at the same time. Indeed, the vector $\mathbf{y}$ in (F2) determines a linear combination of the equations witnessing that $A\mathbf{x} = \mathbf{b}$ cannot have any nonnegative solution: All coefficients on the left-hand side of the resulting equation $(\mathbf{y}^T A)\mathbf{x} = \mathbf{y}^T \mathbf{b}$ are nonnegative, but the right-hand side is negative.

The Farkas lemma is not exactly a difficult theorem, but it is not trivial either. Many proofs are known, and we will present some of them in the subsequent sections. The reader is invited to choose the "best" one according to personal taste.

A geometric view. In order to view the Farkas lemma geometrically, we need the notion of convex hull; see Section 4.3. Further we define, for vectors $\mathbf{a}_1, \mathbf{a}_2, \ldots, \mathbf{a}_n \in \mathbb{R}^m$, the **convex cone** generated by $\mathbf{a}_1, \mathbf{a}_2, \ldots, \mathbf{a}_n$ as the set of all linear combinations of the $\mathbf{a}_i$ with nonnegative coefficients, that is, as

$$\Big\{ t_1 \mathbf{a}_1 + t_2 \mathbf{a}_2 + \cdots + t_n \mathbf{a}_n : t_1, t_2, \ldots, t_n \geq 0 \Big\}.$$

In other words, this convex cone is the convex hull of the rays $p_1, p_2, \ldots, p_n$, where $p_i = \{t\mathbf{a}_i : t \geq 0\}$ emanates from the origin and passes through the point $\mathbf{a}_i$.

6.4.2 Proposition (Farkas lemma geometrically). *Let $\mathbf{a}_1, \mathbf{a}_2, \ldots, \mathbf{a}_n, \mathbf{b}$ be vectors in $\mathbb{R}^m$. Then exactly one of the following two possibilities occurs:*

(F1′) *The point $\mathbf{b}$ lies in the convex cone C generated by $\mathbf{a}_1, \mathbf{a}_2, \ldots, \mathbf{a}_n$.*

(F2′) *There exists a hyperplane h passing through the point $\mathbf{0}$, of the form*

$$h = \{\mathbf{x} \in \mathbb{R}^m : \mathbf{y}^T\mathbf{x} = 0\}$$

for a suitable $\mathbf{y} \in \mathbb{R}^m$, such that all the vectors $\mathbf{a}_1, \mathbf{a}_2, \ldots, \mathbf{a}_n$ (and thus the whole cone C) lie on one side and $\mathbf{b}$ lies (strictly) on the other side. That is, $\mathbf{y}^T\mathbf{a}_i \geq 0$ for all $i = 1, 2, \ldots, n$ and $\mathbf{y}^T\mathbf{b} < 0$.

A drawing illustrates both possibilities for $m = 2$ and $n = 3$:

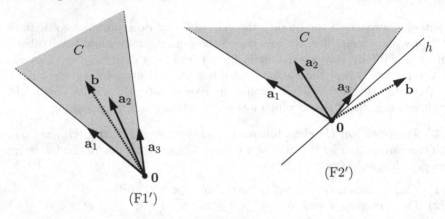

(F1′)   (F2′)

To see that Proposition 6.4.1 and Proposition 6.4.2 really tell us the same thing, it suffices to take the columns of the matrix A for $\mathbf{a}_1, \mathbf{a}_2, \ldots, \mathbf{a}_n$. The existence of a nonnegative solution of $A\mathbf{x} = \mathbf{b}$ can be reexpressed as $\mathbf{b} = t_1\mathbf{a}_1 + t_2\mathbf{a}_2 + \cdots + t_n\mathbf{a}_n, t_1, t_2, \ldots, t_n \geq 0$, and this says exactly that $\mathbf{b} \in C$. The equivalence of (F2) and (F2′) hopefully doesn't need any further explanation.

This result is an instance of a *separation theorem* for convex sets. Separation theorems generally assert that disjoint convex sets can be separated by a hyperplane. There are several versions (depending on whether one requires strict or nonstrict separation, etc.) and several proof strategies. Separation theorems in infinite-dimensional Banach spaces are closely related to the Hahn–Banach theorem, one of the cornerstones of functional analysis. In Section 6.5 we prove the Farkas lemma along these lines, viewing it as a geometric separation theorem.

Variants of the Farkas lemma. Proposition 6.4.1 provides an answer to the question, "When does a system of linear *equalities* have a *nonnegative* solution?" In part (i) of the following proposition, we restate Proposition 6.4.1 (in a slightly different, but clearly equivalent form), and in parts (ii) and (iii), we add two more variants of the Farkas lemma. Part (ii) answers the question, "When does a system of linear *inequalities* have a *nonnegative* solution?" and part (iii) the question, "When does a system of linear *inequalities* have *any* solution at all?"

6.4.3 Proposition (Farkas lemma in three variants). *Let A be a real matrix with m rows and n columns, and let $\mathbf{b} \in \mathbb{R}^m$ be a vector.*

(i) *The system* $A\mathbf{x} = \mathbf{b}$ *has a nonnegative solution if and only if every* $\mathbf{y} \in \mathbb{R}^m$ *with* $\mathbf{y}^T A \geq \mathbf{0}^T$ *also satisfies* $\mathbf{y}^T \mathbf{b} \geq \mathbf{0}$.
(ii) *The system* $A\mathbf{x} \leq \mathbf{b}$ *has a nonnegative solution if and only if every nonnegative* $\mathbf{y} \in \mathbb{R}^m$ *with* $\mathbf{y}^T A \geq \mathbf{0}^T$ *also satisfies* $\mathbf{y}^T \mathbf{b} \geq 0$.
(iii) *The system* $A\mathbf{x} \leq \mathbf{b}$ *has a solution if and only if every nonnegative* $\mathbf{y} \in \mathbb{R}^m$ *with* $\mathbf{y}^T A = \mathbf{0}^T$ *also satisfies* $\mathbf{y}^T \mathbf{b} \geq 0$.

The three parts of Proposition 6.4.3 are mutually equivalent, in the sense that any of them can easily be derived from any other. Having three forms at our disposal provides more flexibility, both for applying the Farkas lemma and for proving it.

The proof of the equivalence (i)⇔(ii)⇔(iii) is easy, using the tricks familiar from transformations of linear programs to equational form. We will take a utilitarian approach: Since we will use (ii) in the proof of the duality theorem, we prove only the implications (i)⇒(ii) and (iii)⇒(ii), leaving the remaining implications to the reader.

Proof of (i)⇒(ii). In (ii) we need an equivalent condition for $A\mathbf{x} \leq \mathbf{b}$ having a nonnegative solution. To this end, we form the matrix $\bar{A} = (A \,|\, I_m)$. We note that $A\mathbf{x} \leq \mathbf{b}$ has a nonnegative solution if and only if $\bar{A}\bar{\mathbf{x}} = \mathbf{b}$ has a nonnegative solution. By (i), this is equivalent to the condition that all $\mathbf{y}$ with $\mathbf{y}^T \bar{A} \geq \mathbf{0}^T$ satisfy $\mathbf{y}^T \mathbf{b} \geq 0$. And finally, $\mathbf{y}^T \bar{A} \geq \mathbf{0}^T$ says exactly the same as $\mathbf{y}^T A \geq \mathbf{0}^T$ and $\mathbf{y} \geq \mathbf{0}$, and hence we have the desired equivalence. □

Proof of (iii)⇒(ii). Again we need an equivalent condition for $A\mathbf{x} \leq \mathbf{b}$ having a nonnegative solution. This time we form the matrix $\bar{A}$ and the vector $\bar{\mathbf{b}}$ according to

$$\bar{A} = \left(\frac{A}{-I_n}\right), \quad \bar{\mathbf{b}} = \left(\frac{\mathbf{b}}{\mathbf{0}}\right).$$

Then $A\mathbf{x} \leq \mathbf{b}$ has a nonnegative solution if and only if $\bar{A}\mathbf{x} \leq \bar{\mathbf{b}}$ has any solution. The latter is equivalent, by (iii), to the condition that all $\bar{\mathbf{y}} \geq \mathbf{0}$ with $\bar{\mathbf{y}}^T \bar{A} = \mathbf{0}^T$ satisfy $\bar{\mathbf{y}}^T \bar{\mathbf{b}} \geq 0$. Writing

$$\bar{\mathbf{y}} = \left(\frac{\mathbf{y}}{\mathbf{y}'}\right),$$

$\mathbf{y}$ a vector with m components, we have

$$\bar{\mathbf{y}} \geq \mathbf{0}, \;\; \bar{\mathbf{y}}^T \bar{A} = \mathbf{0}^T \quad \text{exactly if} \quad \mathbf{y} \geq \mathbf{0}, \;\; \mathbf{y}'^T = \mathbf{y}^T A \geq \mathbf{0}^T$$

and

$$\bar{\mathbf{y}}^T \bar{\mathbf{b}} = \mathbf{y}^T \mathbf{b}.$$

From this and our chain of equivalences, we deduce that $A\mathbf{x} \leq \mathbf{b}$ has a nonnegative solution if and only if all $\mathbf{y} \geq \mathbf{0}$ with $\mathbf{y}^T A \geq \mathbf{0}^T$ satisfy $\mathbf{y}^T \mathbf{b} \geq 0$, and this is the statement of (ii). □

Remarks. A reader with a systematic mind may like to see the variants of the Farkas lemma summarized in a table:

	The system $Ax \le b$	The system $Ax = b$
has a solution $\mathbf{x} \ge \mathbf{0}$ iff	$\mathbf{y} \ge \mathbf{0}, \mathbf{y}^T A \ge \mathbf{0}$ $\Rightarrow \mathbf{y}^T \mathbf{b} \ge 0$	$\mathbf{y}^T A \ge \mathbf{0}^T$ $\Rightarrow \mathbf{y}^T \mathbf{b} \ge 0$
has a solution $\mathbf{x} \in \mathbb{R}^n$ iff	$\mathbf{y} \ge \mathbf{0}, \mathbf{y}^T A = \mathbf{0}$ $\Rightarrow \mathbf{y}^T \mathbf{b} \ge 0$	$\mathbf{y}^T A = \mathbf{0}^T$ $\Rightarrow \mathbf{y}^T \mathbf{b} = 0$

We had three variants of the Farkas lemma, but the table has four entries. We haven't mentioned the statement corresponding to the bottom right corner of the table, telling us when a system of linear *equations* has *any* solution. We haven't mentioned it because it doesn't deserve to be called a Farkas lemma—the proof is a simple exercise in linear algebra, and there doesn't seem to be any way of deriving the Farkas lemma from this variant along the lines of our previous reductions. However, we will find this statement useful in Section 6.6, where it will serve as a basis of a proof of a "real" Farkas lemma.

Let us note that, similar to "dualization for everyone," we could also establish a unifying "Farkas lemma for everyone," dealing with a system containing both linear equations and inequalities and with some of the variables nonnegative and some unrestricted. This would contain all of the four variants considered above as special cases, but we will not go in this direction.

A logical view. Now we explain yet another way of understanding the Farkas lemma, this time variant (iii) in Proposition 6.4.3. We begin with something seemingly different, namely, deriving new linear inequalities from old ones. From two given inequalities, say

$$4x_1 + x_2 \le 4 \quad \text{and} \quad -x_1 + x_2 \le 1,$$

we can derive new inequalities by multiplying the first inequality by a *positive* real number α, the second one by a *positive* real number β, and adding the resulting inequalities together (we must be careful so that both inequality signs have the same direction!); we have already used this many times. For instance, for $\alpha = 3$ and $\beta = 2$ we derive the inequality $10x_1 + 5x_2 \le 14$. More generally, if we start with a system of several linear inequalities, of the form $Ax \le b$, we can derive new inequalities by repeating this operation for various pairs, which may involve both the original inequalities and new ones derived earlier. So if we start with the system

$$4x_1 + x_2 \le 4, \quad -x_1 + x_2 \le 1, \quad \text{and} \quad -2x_1 - x_2 \le -3,$$

we can first derive $10x_1 + 5x_2 \le 14$ from the first two as before, and then we can add to this new inequality the third inequality multiplied by 5. In this case both of the coefficients on the left-hand side cancel out, and we get the inequality $0 \leq -1$. This last inequality *obviously* never holds, and so the original triple of inequalities cannot be satisfied by any $(x_1, x_2) \in \mathbb{R}$ either (as is easy to check using a picture).

The Farkas lemma turns out to be equivalent to the following statement:

> Whenever a system $A\mathbf{x} \leq \mathbf{b}$ of finitely many linear inequalities is **inconsistent**, that is, there is no $\mathbf{x} \in \mathbb{R}^n$ satisfying it, we can derive the (obviously inconsistent) inequality $0 \leq -1$ from it by the above procedure.

A little thought reveals that each inequality derived by the procedure (repeated combinations of pairs) has the form $(\mathbf{y}^T A)\mathbf{x} \leq \mathbf{y}^T \mathbf{b}$ for some nonnegative vector $\mathbf{y} \in \mathbb{R}^m$, and thus, equivalently, we claim that whenever $A\mathbf{x} \leq \mathbf{b}$ is inconsistent, there exists a vector $\mathbf{y} \geq \mathbf{0}$ with $\mathbf{y}^T A = \mathbf{0}^T$ and $\mathbf{y}^T \mathbf{b} = -1$. This is clearly equivalent to part (iii) of Proposition 6.4.3.

The reader may wonder why we have bothered to consider repeated pairwise combinations of inequalities, instead of using a single vector $\mathbf{y}$ specifying a combination of all of the inequalities right away. The reason is that the "pairwise" formulation makes the statement more similar to a number of important and famous statements in various branches of mathematics. In logic, for example, theorems are derived (proved) from axioms by repeated application of certain simple derivation rules. In the first-order propositional calculus, there is a *completeness theorem*: Any true statement (that is, a statement valid in every model) can be derived from the axioms by a finite sequence of steps, using the appropriate derivation rules, such as *modus ponens*. In contrast, the celebrated Gödel's first incompleteness theorem asserts that in Peano arithmetic, as well as in any theory containing it, there are statements that are true but *cannot* be derived.

In analogy to this, we can view the inequalities of the original system $A\mathbf{x} \leq \mathbf{b}$ as "axioms," and we have a single derivation rule (derive a new inequality from two existing ones by a positive linear combination as above). Then the Farkas lemma tells us that any inconsistent system of "axioms" can be refuted by a suitable derivation. (This is a "weak" completeness theorem; we could also consider a more general "completeness theorem," stating that whenever a linear inequality is valid for all $\mathbf{x} \in \mathbb{R}^n$ satisfying $A\mathbf{x} \leq \mathbf{b}$, then it can be derived from $A\mathbf{x} \leq \mathbf{b}$, but we will not go into this here.) Such a completeness result means that the theory of linear inequalities is, in a sense, "easy." Moreover, the simplex method, or also the Fourier–Motzkin elimination considered in Section 6.7, provide ways to construct such a derivation.

This view makes the Farkas lemma a (small) cousin of various completeness theorems of logic and of other famous results, such as

Hilbert's Nullstellensatz in algebraic geometry. Computer science also frequently investigates the possibility of deriving some object from given initial objects by certain derivation rules, say in the theory of formal languages.

Proof of the duality theorem from the Farkas lemma. Let us assume that the linear program (P) has an optimal solution $\mathbf{x}^*$. As in the proof of the duality theorem from the simplex method, we show that the dual (D) has an optimal solution as well, and that the optimum values of both programs coincide.

We first define $\gamma = \mathbf{c}^T\mathbf{x}^*$ to be the optimum value of (P). Then we know that the system of inequalities

$$A\mathbf{x} \le \mathbf{b}, \ \mathbf{c}^T\mathbf{x} \ge \gamma \tag{6.4}$$

has a nonnegative solution, but for any $\varepsilon > 0$, the system

$$A\mathbf{x} \le \mathbf{b}, \ \mathbf{c}^T\mathbf{x} \ge \gamma + \varepsilon \tag{6.5}$$

has *no* nonnegative solution. If we define an $(m+1) \times n$ matrix $\hat{A}$ and a vector $\hat{\mathbf{b}}_\varepsilon \in \mathbb{R}^m$ by

$$\hat{A} = \begin{pmatrix} A \\ -\mathbf{c}^T \end{pmatrix}, \quad \hat{\mathbf{b}}_\varepsilon = \begin{pmatrix} \mathbf{b} \\ -\gamma - \varepsilon \end{pmatrix},$$

then (6.4) is equivalent to $\hat{A}\mathbf{x} \le \hat{\mathbf{b}}_0$ and (6.5) is equivalent to $\hat{A}\mathbf{x} \le \hat{\mathbf{b}}_\varepsilon$.

Let us apply the variant of the Farkas lemma in Proposition 6.4.3(ii). For $\varepsilon > 0$, the system $\hat{A}\mathbf{x} \le \hat{\mathbf{b}}_\varepsilon$ has no nonnegative solution, so we conclude that there is a nonnegative vector $\hat{\mathbf{y}} = (\mathbf{u}, z) \in \mathbb{R}^{m+1}$ such that $\hat{\mathbf{y}}^T \hat{A} \ge \mathbf{0}^T$ but $\hat{\mathbf{y}}^T \hat{\mathbf{b}}_\varepsilon < 0$. These conditions boil down to

$$A^T\mathbf{u} \ge z\mathbf{c}, \ \mathbf{b}^T\mathbf{u} < z(\gamma + \varepsilon). \tag{6.6}$$

Applying the Farkas lemma in the case $\varepsilon = 0$ (the system has a nonnegative solution), we see that the very same vector $\hat{\mathbf{y}}$ must satisfy $\hat{\mathbf{y}}^T\hat{\mathbf{b}}_0 \ge 0$, and this is equivalent to

$$\mathbf{b}^T\mathbf{u} \ge z\gamma.$$

It follows that $z > 0$, since $z = 0$ would contradict the strict inequality in (6.6). But then we may set $\mathbf{v} := \frac{1}{z}\mathbf{u} \ge \mathbf{0}$, and (6.6) yields

$$A^T\mathbf{v} \ge \mathbf{c}, \ \mathbf{b}^T\mathbf{v} < \gamma + \varepsilon.$$

In other words, $\mathbf{v}$ is a feasible solution of (D), with the value of the objective function smaller than $\gamma + \varepsilon$.

By the weak duality theorem, every feasible solution of (D) has value of the objective function at least γ. Hence (D) is a feasible and bounded linear program, and so we know that it has an optimal solution $\mathbf{y}^*$ (Theorem 4.2.3). Its value $\mathbf{b}^T\mathbf{y}^*$ is between γ and $\gamma + \varepsilon$ for every $\varepsilon > 0$, and thus it equals γ. This concludes the proof of the duality theorem. $\square$

6.5 Farkas Lemma: An Analytic Proof

In this section we prove the geometric version of the Farkas lemma, Proposition 6.4.2, by means of elementary geometry and analysis. We are given vectors $\mathbf{a}_1, \ldots, \mathbf{a}_n$ in $\mathbb{R}^m$, and we let C be the convex cone generated by them, i.e., the set of all linear combinations with nonnegative coefficients. Proving the Farkas lemma amounts to showing that for any vector $\mathbf{b} \notin C$ there exists a hyperplane separating it from C and passing through $\mathbf{0}$. In other words, we want to exhibit a vector $\mathbf{y} \in \mathbb{R}^m$ with $\mathbf{y}^T \mathbf{b} < 0$ and $\mathbf{y}^T \mathbf{x} \geq 0$ for all $\mathbf{x} \in C$.

The plan of the proof is straightforward: We let $\mathbf{z}$ be the point of C nearest to $\mathbf{b}$ (in the Euclidean distance), and we check that the vector $\mathbf{y} = \mathbf{z} - \mathbf{b}$ is as required; see the following illustration:

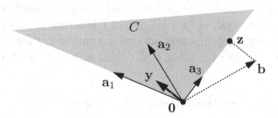

The main technical part of the proof is to show that the nearest point $\mathbf{z}$ exists. Indeed, in principle, it might happen that no point is the nearest (for example, such a situation occurs for the point 0 on the real line and the open interval $(1, 2)$; the interval contains points with distance to 0 as close to 1 as desired, but no point at distance exactly 1).

6.5.1 Lemma. *Let C be a convex cone in $\mathbb{R}^m$ generated by finitely many vectors $\mathbf{a}_1, \ldots, \mathbf{a}_n$, and let $\mathbf{b} \notin C$ be a point. Then there exists a point $\mathbf{z} \in C$ nearest to $\mathbf{b}$ (it is also unique but we won't need this).*

Proof of Proposition 6.4.2 assuming Lemma 6.5.1. As announced, we set $\mathbf{y} = \mathbf{z} - \mathbf{b}$, where $\mathbf{z}$ is a point of C nearest to $\mathbf{b}$.

First we check that $\mathbf{y}^T \mathbf{z} = 0$. This is clear for $\mathbf{z} = \mathbf{0}$. For $\mathbf{z} \neq \mathbf{0}$, if $\mathbf{z}$ were not perpendicular to $\mathbf{y}$, we could move $\mathbf{z}$ slightly along the ray $\{t\mathbf{z} : t \geq 0\} \subseteq C$ and get a point closer to $\mathbf{b}$. More formally, let us first assume that $\mathbf{y}^T \mathbf{z} > 0$, and let us set $\mathbf{z}' = (1 - \alpha)\mathbf{z}$ for a small $\alpha > 0$. We calculate $\|\mathbf{z}' - \mathbf{b}\|^2 = (\mathbf{y} - \alpha\mathbf{z})^T(\mathbf{y} - \alpha\mathbf{z}) = \|\mathbf{y}\|^2 - 2\alpha\mathbf{y}^T\mathbf{z} + \alpha^2\|\mathbf{z}\|^2$. We have $2\alpha\mathbf{y}^T\mathbf{z} > \alpha^2\|\mathbf{z}\|^2$ for all sufficiently small $\alpha > 0$, and thus $\|\mathbf{z}' - \mathbf{b}\|^2 < \|\mathbf{y}\|^2 = \|\mathbf{z} - \mathbf{b}\|^2$. This contradicts $\mathbf{z}$ being a nearest point. The case $\mathbf{y}^T \mathbf{z} < 0$ is handled similarly.

To verify $\mathbf{y}^T \mathbf{b} < 0$, we recall that $\mathbf{y} \neq \mathbf{0}$, and we compute $0 < \mathbf{y}^T\mathbf{y} = \mathbf{y}^T\mathbf{z} - \mathbf{y}^T\mathbf{b} = -\mathbf{y}^T\mathbf{b}$.

Next, let $\mathbf{x} \in C$, $\mathbf{x} \neq \mathbf{z}$. The angle $\angle\mathbf{bzx}$ has to be at least 90 degrees, for otherwise, points on the segment $\mathbf{zx}$ sufficiently close to $\mathbf{z}$ would lie closer to $\mathbf{b}$ than $\mathbf{z}$; equivalently, $(\mathbf{b} - \mathbf{z})^T(\mathbf{x} - \mathbf{z}) \leq 0$ (this is similar to the above argument for $\mathbf{y}^T\mathbf{z} = 0$ and we leave a formal verification to the reader). Thus $0 \geq (\mathbf{b} - \mathbf{z})^T(\mathbf{x} - \mathbf{z}) = -\mathbf{y}^T\mathbf{x} + \mathbf{y}^T\mathbf{z} = -\mathbf{y}^T\mathbf{x}$. The Farkas lemma is proved. $\quad\square$

It remains to prove Lemma 6.5.1. We do it in several steps, and each of them is an interesting little fact in itself.

6.5.2 Lemma. *Let $X \subseteq \mathbb{R}^m$ be a nonempty closed set and let $\mathbf{b} \in \mathbb{R}^m$ be a point. Then X has (at least one) point nearest to $\mathbf{b}$.*

Proof. This is simple but it needs basic facts about compact sets in $\mathbb{R}^d$. Let us fix an arbitrary $\mathbf{x}_0 \in X$, let $r = \|\mathbf{x}_0 - \mathbf{b}\|$, and let $K = \{\mathbf{x} \in X : \|\mathbf{x} - \mathbf{b}\| \leq r\}$. Clearly, if K has a nearest point to $\mathbf{b}$, then the same point is a point of X nearest to $\mathbf{b}$. Since K is the intersection of X with a closed ball of radius r, it is closed and bounded, and hence compact. We define the function $f: K \to \mathbb{R}$ by $f(\mathbf{x}) = \|\mathbf{x} - \mathbf{b}\|$. Then f is a continuous function on a compact set, and any such function attains a minimum; that is, there exists $\mathbf{z} \in K$ with $f(\mathbf{z}) \leq f(\mathbf{x})$ for all $\mathbf{x} \in K$. Such a $\mathbf{z}$ is a point of K nearest to $\mathbf{b}$. $\quad\square$

So it remains to prove the following statement:

6.5.3 Lemma. *Every finitely generated convex cone is closed.*

This lemma is not as obvious as it might seem. As a warning example, let us consider a closed disk D in the plane with $\mathbf{0}$ on the boundary. Then the cone generated by D, that is, the set $\{t\mathbf{x} : \mathbf{x} \in D\}$, is an open half-plane plus the point $\mathbf{0}$, and thus it is not closed. Of course, this doesn't contradict to the lemma, but it shows that we must use the finiteness somehow.

Let us define a *primitive cone* in $\mathbb{R}^m$ as a convex cone generated by some $k \leq m$ linearly independent vectors. Before proving Lemma 6.5.3, we deal with the following special case:

6.5.4 Lemma. *Every primitive cone P in $\mathbb{R}^m$ is closed.*

Proof. Let $P_0 \subseteq \mathbb{R}^k$ be the cone generated by the vectors $\mathbf{e}_1, \ldots, \mathbf{e}_k$ of the standard basis of $\mathbb{R}^k$. In other words, P_0 is the nonnegative orthant, and its closedness is hopefully beyond any doubt (for example, it is the intersection of the closed half-spaces $x_i \geq 0, i = 1, 2, \ldots, k$).

Let the given primitive cone $P \subseteq \mathbb{R}^m$ be generated by linearly independent vectors $\mathbf{a}_1, \ldots, \mathbf{a}_k$. We define a linear mapping $f: \mathbb{R}^k \to \mathbb{R}^m$ by $f(\mathbf{x}) = x_1\mathbf{a}_1 + x_2\mathbf{a}_2 + \cdots + x_k\mathbf{a}_k$. This f is injective by the linear independence of the $\mathbf{a}_j$, and we have $P = f(P_0)$. So it suffices to prove the following claim: *The image $P = f(P_0)$ of a closed set P_0 under an injective linear map $f: \mathbb{R}^k \to \mathbb{R}^m$ is closed.*

To see this, we let $L = f(\mathbb{R}^k)$ be the image of f. Since f is injective, it is a linear isomorphism of $\mathbb{R}^k$ and L. A linear isomorphism f has a linear inverse map $g = f^{-1} \colon L \to \mathbb{R}^k$. Every linear map between Euclidean spaces is continuous (this can be checked using a matrix form of the map), and we have $P = g^{-1}(P_0)$. The preimage of a closed set under a continuous map is closed by definition (while the *image* of a closed set under a continuous map need not be closed in general!), so P is closed as a subset of L. Since L is closed in $\mathbb{R}^m$ (being a linear subspace), we get that P is closed as desired. □

Lemma 6.5.3 is now a consequence of Lemma 6.5.4, of the fact that the union of finitely many closed sets is closed, and of the next lemma:

6.5.5 Lemma. *Let C be a convex cone in $\mathbb{R}^m$ generated by finitely many vectors $\mathbf{a}_1, \ldots, \mathbf{a}_n$. Then C can be expressed as a union of finitely many primitive cones.*

Proof. For every $\mathbf{x} \in C$ we are going to verify that it is contained in a primitive cone generated by a suitable set of linearly independent vectors among the $\mathbf{a}_i$. We may assume $\mathbf{x} \neq \mathbf{0}$ (since $\{\mathbf{0}\}$ is the primitive cone generated by the empty set of vectors).

Let $I \subseteq \{1, 2, \ldots, n\}$ be a set of minimum possible size such that $\mathbf{x}$ lies in the convex cone generated by $A_I = \{\mathbf{a}_i : i \in I\}$ (this is a standard trick in linear algebra and in convex geometry). That is, there exist nonnegative coefficients α_i, $i \in I$, with $\mathbf{x} = \sum_{i \in I} \alpha_i \mathbf{a}_i$. The α_i are even strictly positive since if some $\alpha_i = 0$, we could delete i from I. We now want to show that the set A_I is linearly independent. For contradiction, we suppose that there is a nontrivial linear combination $\sum_{i \in I} \beta_i \mathbf{a}_i = \mathbf{0}$, where not all β_i are 0. Then there exists a real t such that all the expressions $\alpha_i - t\beta_i$ are nonnegative and at least one of them is zero. (To see this, we can first consider the case that some β_i is strictly positive, we start with $t = 0$, we let it grow, and see what happens. The case of a strictly negative β_i is analogous with t decreasing from the initial value 0.) Then the equation

$$\mathbf{x} = \sum_{i \in I} (\alpha_i - t\beta_i)\mathbf{a}_i$$

expresses $\mathbf{x}$ as a linear combination with positive coefficients of fewer than $|I|$ vectors. □

6.6 Farkas Lemma from Minimally Infeasible Systems

Here we derive the Farkas lemma from an observation concerning minimally infeasible systems. A system $Ax \leq b$ of m inequalities is called *minimally infeasible* if the system has no solution, but every subsystem obtained by dropping one inequality does have a solution.

6.6.1 Lemma. *Let $A\mathbf{x} \leq \mathbf{b}$ be a minimally infeasible system of m inequalities, and let $A^{(i)}\mathbf{x} \leq \mathbf{b}^{(i)}$ be the subsystem obtained by dropping the ith inequality, $i = 1, 2, \ldots, m$. Then for every i there exists a vector $\tilde{\mathbf{x}}^{(i)}$ such that $A^{(i)}\tilde{\mathbf{x}}^{(i)} = \mathbf{b}^{(i)}$.*

Let us set $\mathbf{a}_i = (a_{i1}, a_{i2}, \ldots, a_{in})$ and write the ith inequality as $\mathbf{a}_i^T \mathbf{x} \leq b_i$. Here is an illustration for an example in the plane ($n = 2$) with $m = 3$ inequalities:

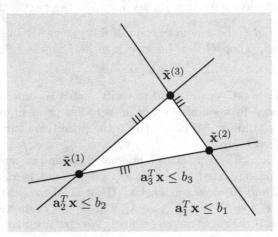

Proof. We consider the linear program

$$\begin{aligned} \text{minimize} \quad & z \\ \text{subject to} \quad & A\mathbf{x} \leq \mathbf{b} + z\mathbf{e}_i, \end{aligned} \qquad (\text{LP}^{(i)})$$

where $\mathbf{e}_i$ is the ith unit vector. The idea of $(\text{LP}^{(i)})$ is to translate the half-space $\{\mathbf{x} : \mathbf{a}_i^T \mathbf{x} \leq b_i\}$ by the minimum amount necessary to achieve feasibility. For the example illustrated above and $i = 3$, this results in the following picture:

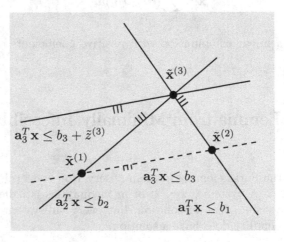

To show formally that (LP$^{(i)}$) has an optimal solution, we first argue that it has a feasible solution. Indeed, by the assumption, the system $A^{(i)}\mathbf{x} \leq \mathbf{b}^{(i)}$ has at least one solution. Let us fix an arbitrary solution of this system and denote it by $\bar{\mathbf{x}}$. We put $\bar{z} = \mathbf{a}_i^T \bar{\mathbf{x}} - b_i$, and we note that the vector $(\bar{\mathbf{x}}, \bar{z})$ is a feasible solution of the linear program (LP$^{(i)}$).

Next, we note that (LP$^{(i)}$) is also bounded, since $A\mathbf{x} \leq \mathbf{b}$ has no solution. Therefore, the linear program has an optimal solution $(\tilde{\mathbf{x}}^{(i)}, \tilde{z}^{(i)})$ with $\tilde{z}^{(i)} > 0$ by Theorem 4.2.3.

We claim that the just defined $\tilde{\mathbf{x}}^{(i)}$ satisfies $A^{(i)}\tilde{\mathbf{x}}^{(i)} = \mathbf{b}^{(i)}$. We already know that $A^{(i)}\tilde{\mathbf{x}}^{(i)} \leq \mathbf{b}^{(i)}$. Let us suppose for contradiction that $\mathbf{a}_j^T \tilde{\mathbf{x}}^{(i)} = b_j - \varepsilon$ for some $j \neq i$ and $\varepsilon > 0$. We will show that then $(\tilde{\mathbf{x}}^{(i)}, \tilde{z}^{(i)})$ cannot be optimal for (LP$^{(i)}$). To this end, let us consider an optimal solution $(\tilde{\mathbf{x}}^{(j)}, \tilde{z}^{(j)})$ of (LP$^{(j)}$). The idea is that by moving the point $(\tilde{\mathbf{x}}^{(i)}, \tilde{z}^{(i)})$ slightly towards $(\tilde{\mathbf{x}}^{(j)}, \tilde{z}^{(j)})$, we remain feasible for (LP$^{(i)}$), but we improve the objective function of (LP$^{(i)}$). More formally, for a real number $t \geq 0$, we define $\tilde{\mathbf{x}}(t) = (1 - t)\tilde{\mathbf{x}}^{(i)} + t\tilde{\mathbf{x}}^{(j)}$. It follows that

$$\mathbf{a}_j^T \tilde{\mathbf{x}}(t) \leq b_j - (1 - t)\varepsilon + t\tilde{z}^{(j)},$$
$$\mathbf{a}_i^T \tilde{\mathbf{x}}(t) \leq b_i + (1 - t)\tilde{z}^{(i)},$$
$$\mathbf{a}_k^T \tilde{\mathbf{x}}(t) \leq b_k, \text{ for all } k \neq i, j.$$

Thus for t sufficiently small, namely, for $0 < t \leq (1 - t)\varepsilon/\tilde{z}^{(j)}$, the pair $(\tilde{\mathbf{x}}(t), (1-t)\tilde{z}^{(i)})$ is a feasible solution of (LP$^{(i)}$) with objective function strictly smaller than $\tilde{z}^{(i)}$, contradicting the assumed optimality of $(\tilde{\mathbf{x}}^{(i)}, \tilde{z}^{(i)})$. Thus, $A^{(i)}\tilde{\mathbf{x}}^{(i)} = \mathbf{b}^{(i)}$ and the lemma is proved.   □

We need another lemma that proves an "easy" variant of the Farkas lemma, concerned with *arbitrary* solutions of systems of *equalities*.

This lemma establishes the implication in the bottom right corner of the table of Farkas lemma variants on page 92.

6.6.2 Lemma. *The system $A\mathbf{x} = \mathbf{b}$ has a solution if and only if every $\mathbf{y} \in \mathbb{R}^m$ with $\mathbf{y}^T A = \mathbf{0}^T$ also satisfies $\mathbf{y}^T \mathbf{b} = 0$.*

Proof. One direction is easy. If $A\mathbf{x} = \mathbf{b}$ has some solution $\tilde{\mathbf{x}}$, and if $\mathbf{y}^T A = \mathbf{0}^T$, then $0 = \mathbf{0}^T \tilde{\mathbf{x}} = \mathbf{y}^T A\tilde{\mathbf{x}} = \mathbf{y}^T \mathbf{b}$.

If $A\mathbf{x} = \mathbf{b}$ has no solution, we need to find a vector $\mathbf{y}$ such that $\mathbf{y}^T A = \mathbf{0}^T$ and $\mathbf{y}^T \mathbf{b} \neq 0$. Let us define $r = \text{rank}(A)$ and consider the $m \times (n + 1)$ matrix $(A \,|\, \mathbf{b})$. This matrix has rank $r + 1$ since the last column is not a linear combination of the first n columns. For the very same reason, the matrix

$$\left(\begin{array}{c|c} A & \mathbf{b} \\ \hline \mathbf{0}^T & -1 \end{array} \right)$$

has rank $r+1$. This shows that the row vector $(\mathbf{0}^T \mid -1)$ is a linear combination of rows of $(A \mid \mathbf{b})$, and the coefficients of this linear combination define a vector $\mathbf{y} \in \mathbb{R}^m$ with $\mathbf{y}^T A = \mathbf{0}^T$ and $\mathbf{y}^T \mathbf{b} = -1$, as desired.   □

Now we proceed to the proof of the Farkas lemma. The variant that results most naturally is the one with an arbitrary solution of $A\mathbf{x} \leq \mathbf{b}$, that is, Proposition 6.4.3(iii).

Proof of Proposition 6.4.3(iii). As in Lemma 6.6.2, one direction is easy: If $A\mathbf{x} \leq \mathbf{b}$ has some solution $\tilde{\mathbf{x}}$, and if $\mathbf{y} \geq \mathbf{0}$, $\mathbf{y}^T A = \mathbf{0}^T$, we get $0 = \mathbf{0}^T \tilde{\mathbf{x}} = \mathbf{y}^T A \tilde{\mathbf{x}} \leq \mathbf{y}^T \mathbf{b}$. The interesting case is that $A\mathbf{x} \leq \mathbf{b}$ has no solution. Our task is then to construct a vector $\mathbf{y} \geq \mathbf{0}$ satisfying $\mathbf{y}^T A = \mathbf{0}^T$ and $\mathbf{y}^T \mathbf{b} < 0$.

We may assume that $A\mathbf{x} \leq \mathbf{b}$ is minimally infeasible, by restricting to a suitable subsystem: A vector $\mathbf{y}$ for this subsystem can be extended to work for the original system by inserting zeros at appropriate places.

Since $A\mathbf{x} \leq \mathbf{b}$ has no solution, the system $A\mathbf{x} = \mathbf{b}$ has no solution either. By Lemma 6.6.2, there exists a vector $\mathbf{y} \in \mathbb{R}^m$ such that $\mathbf{y}^T A = \mathbf{0}^T$ and $\mathbf{y}^T \mathbf{b} \neq 0$. By possibly changing signs, we may assume that $\mathbf{y}^T \mathbf{b} < 0$. We will show that this vector also satisfies $\mathbf{y} \geq \mathbf{0}$, and this will finish the proof. To this end, we fix $i \in \{1, 2, \ldots, m\}$ and consider the vector $\tilde{\mathbf{x}}^{(i)}$ as in Lemma 6.6.1 above. With the terminology of the lemma, we have $A^{(i)} \tilde{\mathbf{x}}^{(i)} = \mathbf{b}^{(i)}$, and using $\mathbf{y}^T A = \mathbf{0}^T$, we can write

$$y_i(\mathbf{a}_i^T \tilde{\mathbf{x}}^{(i)} - b_i) = \mathbf{y}^T(A\tilde{\mathbf{x}}^{(i)} - \mathbf{b}) = -\mathbf{y}^T \mathbf{b} > 0.$$

Proposition 6.4.3(iii) is proved.   □

This proof of the Farkas lemma is based on the paper

> M. Conforti, M. Di Summa, and G. Zambelli: Minimally infeasible set partitioning problems with balanced constraints, *Mathematics of Operations Research*, to appear.

The proof given there is even more elementary than ours in the sense that it does not use linear programming. We have chosen the linear programming approach since we find it somewhat more transparent.

6.7 Farkas Lemma from the Fourier–Motzkin Elimination

When explaining the "logical view" of the Farkas lemma in Section 6.4, we started with a system of 3 inequalities and combined pairs of inequalities together, until we managed to eliminate all variables and obtained the obviously unsatisfiable inequality $0 \leq -1$. The Fourier–Motzkin elimination is a

systematic procedure for eliminating all variables from an arbitrary system
$A\mathbf{x} \le \mathbf{b}$ of linear inequalities. If the final inequalities with no variables hold,
we can reconstruct a solution of the original system by tracing the computa-
tions backward, and if one of the final inequalities does not hold, it certifies
that the original system has no solution.

The Fourier–Motzkin elimination is similar in spirit to Gaussian elimination for systems of linear equations, and it is just as simple. As in Gaussian
elimination, variables are removed one at a time, but there is a price to pay:
To get rid of one variable, we typically have to introduce many new inequalities, so that the method becomes impractical already for moderately large
systems. The Fourier–Motzkin elimination can be considered as a simple but
inefficient alternative to the simplex method. For the purpose of proving
statements about systems of inequalities, efficiency is not a concern, so it is
the simplicity of the Fourier–Motzkin elimination that makes it a very handy
tool.

As an example, let us consider the following system of 5 inequalities in
3 variables:

$$
\begin{aligned}
2x - 5y + 4z &\le 10 \\
3x - 6y + 3z &\le 9 \\
5x + 10y - z &\le 15 \\
-x + 5y - 2z &\le -7 \\
-3x + 2y + 6z &\le 12.
\end{aligned}
\tag{6.7}
$$

In the first step we would like to eliminate x. For a moment let us imagine that
y and z are some fixed real numbers, and let us ask under what conditions
we can choose a value of x such that together with the given values y and z it
satisfies (6.7). The first three inequalities impose an upper bound on x, while
the remaining two impose a lower bound. To make this clearer, we rewrite
the system as follows:

$$
\begin{aligned}
x &\le 5 + \tfrac{5}{2}y - 2z \\
x &\le 3 + 2y - z \\
x &\le 3 - 2y + \tfrac{1}{5}z \\
x &\ge 7 + 5y - 2z \\
x &\ge -4 + \tfrac{2}{3}y + 2z.
\end{aligned}
$$

So given y and z, the admissible values of x are exactly those in the interval
from $\max(7+5y-2z, -4+\tfrac{2}{3}y+2z)$ to $\min(5+\tfrac{5}{2}y-2z, 3+2y-z, 3-2y+\tfrac{1}{5}z)$. If
this interval happens to be empty, there is no admissible x. So the inequality

$$
\begin{aligned}
\max(7 + 5y - 2z, -4 + \tfrac{2}{3}y + 2z) \\
\le \min(5 + \tfrac{5}{2}y - 2z, 3 + 2y - z, 3 - 2y + \tfrac{1}{5}z)
\end{aligned}
\tag{6.8}
$$

is *equivalent* to the existence of x that together with the considered y and z
solves (6.7). The key observation in the Fourier–Motzkin elimination is that
(6.8) can be rewritten as a system of linear inequalities in the variables y
and z. The inequalities simply say that *each of the lower bounds is less than
or equal to each of the upper bounds*:

$$7 + 5y - 2z \leq 5 + \tfrac{5}{2}y - 2z$$
$$7 + 5y - 2z \leq 3 + 2y - z$$
$$7 + 5y - 2z \leq 3 - 2y + \tfrac{1}{5}z$$
$$-4 + \tfrac{2}{3}y + 2z \leq 5 + \tfrac{5}{2}y - 2z$$
$$-4 + \tfrac{2}{3}y + 2z \leq 3 + 2y - z$$
$$-4 + \tfrac{2}{3}y + 2z \leq 3 - 2y + \tfrac{1}{5}z.$$

If we rewrite this system in the usual form $A\mathbf{x} \leq \mathbf{b}$, we arrive at

$$
\begin{aligned}
\tfrac{5}{2}y & & \leq -2 \\
3y & - z & \leq -4 \\
7y & - \tfrac{11}{5}z & \leq -4 \\
-\tfrac{11}{6}y & + 4z & \leq 9 \\
-\tfrac{4}{3}y & + 3z & \leq 7 \\
\tfrac{8}{3}y & + \tfrac{9}{5}z & \leq 7.
\end{aligned}
\tag{6.9}
$$

This system has a solution exactly if the original system (6.7) has one, but it has one variable fewer. The reader is invited to continue with this example, eliminating y and then z. We note that (6.9) gives 4 upper bounds for y and 2 lower bounds, and hence we obtain 8 inequalities after eliminating y.

For larger systems the number of inequalities generated by the Fourier–Motzkin elimination tends to explode. This wasn't so apparent for our small example, but if we have m inequalities and, say, half of them impose upper bounds on the first variable and half impose lower bounds, then we get about $m^2/4$ inequalities after eliminating the first variable, about $m^4/16$ after eliminating the second variable (again, provided that about half of the inequalities give upper bounds for the second variable and half lower bounds), etc.

Now we formulate the procedure in general.

Claim. Let $A\mathbf{x} \leq \mathbf{b}$ be a system with $n \geq 1$ variables and m inequalities. There is a system $A'\mathbf{x}' \leq \mathbf{b}'$ with $n-1$ variables and at most $\max(m, m^2/4)$ inequalities, with the following properties:

(i) $A\mathbf{x} \leq \mathbf{b}$ has a solution if and only if $A'\mathbf{x}' \leq \mathbf{b}'$ has a solution, and
(ii) each inequality of $A'\mathbf{x}' \leq \mathbf{b}'$ is a positive linear combination of some inequalities from $A\mathbf{x} \leq \mathbf{b}$.

Proof. We classify the inequalities into three groups, depending on the coefficient of x_1. We call the ith inequality of $A\mathbf{x} \leq \mathbf{b}$ a *ceiling* if $a_{i1} > 0$, and we call it a *floor* if $a_{i1} < 0$. Otherwise (if $a_{i1} = 0$), it is a *level*. Let $C, F, L \subseteq \{1, \ldots, m\}$ collect the indices of ceilings, floors, and levels. We may assume that

$$
a_{i1} = \begin{cases}
1 & \text{if } i \in C \\
-1 & \text{if } i \in F \\
0 & \text{if } i \in L.
\end{cases}
\tag{6.10}
$$

This situation can be reached by multiplying each inequality in $A\mathbf{x} \leq \mathbf{b}$ by a suitable positive number, which does not change the set of solutions.

Now we can eliminate x_1 between all pairs of ceilings and floors, by simply adding up the two inequalities for each pair.

If $\mathbf{x}'$ is the (possibly empty) vector $(x_2, \ldots, x_n)$, and $\mathbf{a}'_i$ is the (possibly empty) vector $(a_{i2}, \ldots, a_{in})$, then the following inequalities are implied by $A\mathbf{x} \leq \mathbf{b}$:

$$\mathbf{a}'^T_j \mathbf{x}' + \mathbf{a}'^T_k \mathbf{x}' \leq b_j + b_k, \quad j \in C, k \in F. \tag{6.11}$$

The level inequalities of $A\mathbf{x} \leq \mathbf{b}$ can be rewritten as

$$\mathbf{a}'^T_\ell \mathbf{x}' \leq b_\ell, \quad \ell \in L. \tag{6.12}$$

So if $A\mathbf{x} \leq \mathbf{b}$ has a solution, then the system of $|C| \cdot |F| + |L|$ inequalities in $n - 1$ variables given by (6.11) and (6.12) has a solution as well. Conversely, if the latter system has a solution $\tilde{\mathbf{x}}' = (\tilde{x}_2, \ldots, \tilde{x}_n)$, we can determine a suitable value $\tilde{x}_1$ such that the vector $(\tilde{x}_1, \tilde{x}_2, \ldots, \tilde{x}_n)$ solves $A\mathbf{x} \leq \mathbf{b}$. To find $\tilde{x}_1$, we first observe that (6.11) is equivalent to

$$\mathbf{a}'^T_k \mathbf{x}' - b_k \leq b_j - \mathbf{a}'^T_j \mathbf{x}', \quad j \in C, k \in F.$$

This in particular implies

$$\max_{k \in F} \left(\mathbf{a}'^T_k \tilde{\mathbf{x}}' - b_k \right) \leq \min_{j \in C} \left(b_j - \mathbf{a}'^T_j \tilde{\mathbf{x}}' \right).$$

We let $\tilde{x}_1$ be any value between these bounds. It follows that

$$
\begin{aligned}
\tilde{x}_1 + \mathbf{a}'^T_j \tilde{\mathbf{x}}' &\leq b_j, \quad j \in C, \\
-\tilde{x}_1 + \mathbf{a}'^T_k \tilde{\mathbf{x}}' &\leq b_k, \quad k \in F.
\end{aligned}
$$

By our assumption (6.10), we have a feasible solution of the original system $A\mathbf{x} \leq \mathbf{b}$. We note that this argument also works for $C = \emptyset$ or $F = \emptyset$, with the usual convention that $\max_{t \in \emptyset} f(t) = -\infty$ and $\min_{t \in \emptyset} f(t) = \infty$.

Now we can prove the Farkas lemma. The variant that results most naturally from the Fourier–Motzkin elimination is (as in Section 6.6) the one with an arbitrary solution of $A\mathbf{x} \leq \mathbf{b}$, that is, Proposition 6.4.3(iii).

Proof of Proposition 6.4.3(iii). One direction is easy. If $A\mathbf{x} \leq \mathbf{b}$ has some solution $\tilde{\mathbf{x}}$, and $\mathbf{y} \geq \mathbf{0}$ satisfies $\mathbf{y}^T A = \mathbf{0}^T$, we get $0 = \mathbf{0}^T \tilde{\mathbf{x}} = \mathbf{y}^T A \tilde{\mathbf{x}} \leq \mathbf{y}^T \mathbf{b}$. If $A\mathbf{x} \leq \mathbf{b}$ has no solution, then our task is to construct a vector $\mathbf{y}$ satisfying

$$\mathbf{y} \geq \mathbf{0}, \quad \mathbf{y}^T A = \mathbf{0}^T, \quad \text{and } \mathbf{y}^T \mathbf{b} < 0. \tag{6.13}$$

To find such a witness of infeasibility, we use induction on the number of variables. Let us first consider the base case in which the system $Ax \leq b$ has no variables, meaning that it is of the form $\mathbf{0} \leq \mathbf{b}$ with $b_i < 0$ for some i. We set $\mathbf{y} = \mathbf{e}_i$ (the ith unit vector), and this clearly satisfies the requirements for $\mathbf{y}$ being a witness of infeasibility (the condition $\mathbf{y}^T A = \mathbf{0}^T$ is vacuous, since A has no column).

If $A\mathbf{x} \leq \mathbf{b}$ has at least one variable, we perform a step of the Fourier–Motzkin elimination. This yields an infeasible system $A'\mathbf{x}' \leq \mathbf{b}'$, consisting of the inequalities (6.11) and (6.12). Because the latter system has one variable fewer, we inductively find a witness of infeasibility $\mathbf{y}'$ for it. We recall that all inequalities of $A'\mathbf{x}' \leq \mathbf{b}'$ are positive linear combinations of original inequalities; equivalently, there is an $m \times m$ matrix M with all entries nonnegative and

$$(\mathbf{0} \,|\, A') = MA, \quad \mathbf{b}' = M\mathbf{b}.$$

We claim that $\mathbf{y} = M^T\mathbf{y}'$ is a witness of infeasibility for the original system $A\mathbf{x} \leq \mathbf{b}$. Indeed, we have $\mathbf{y}^T A = \mathbf{y}'^T MA = \mathbf{y}'^T (\mathbf{0} \,|\, A') = \mathbf{0}^T$ and $\mathbf{y}^T\mathbf{b} = \mathbf{y}'^T M\mathbf{b} = \mathbf{y}'^T \mathbf{b}' < 0$, since $\mathbf{y}'$ is a witness of infeasibility for $A'\mathbf{x}' \leq \mathbf{b}'$. The condition $\mathbf{y} \geq \mathbf{0}$ follows from $\mathbf{y}' \geq \mathbf{0}$ by the nonnegativity of M.   □

7. Not Only the Simplex Method

Tens of different algorithms have been suggested for linear programming over the years. Most of them didn't work very well, and only very few have turned out as serious competitors to the simplex method, the historically first algorithm. But at least two methods raised great excitement at the time of discovery and they are surely worth mentioning.

The first of them, the *ellipsoid method*, cannot compete with the simplex method in practice, but it had immense theoretical significance. It is the first linear programming algorithm for which it was proved that it always runs in polynomial time (which is not known about the simplex method up to the present, and for many pivot rules it is not even true).

The second is the *interior point method*, or rather, we should say interior point *methods*, since it is an entire group of algorithms. For some of them a polynomial bound on the running time has also been proved, but moreover, these algorithms successfully compete with the simplex method in practice. It seems that for some types of linear programs the simplex method is better, while for others interior point methods are the winners.

Let us remark that several other algorithms, closely related to the simplex method, are used for linear programming as well. The *dual simplex method* can roughly be described as the simplex method applied to the dual linear program. But details of the implementation, which are crucial for the speed of the algorithm in practice, are somewhat different. The dual simplex method is particularly suitable for linear programs that in equational form have $n - m$ significantly smaller than m.

The *primal–dual method* goes through a sequence of feasible solutions of the dual linear program. To get from one such solution to the next, it does not perform a pivot step, but it solves an auxiliary problem that may be derived from the primal linear program or by other means. This greater freedom can be useful, for instance, in approximation algorithms for combinatorial optimization problems.

A little more about the dual simplex method and the primal–dual method can be found in the glossary.

7.1 The Ellipsoid Method

The ellipsoid method was invented in 1970 by Shor, Judin, and Nemirovski as an algorithm for certain nonlinear optimization problems. In 1979 Leonid Khachyian outlined, in a short note, how linear programs can be solved by this method in provably polynomial time. The world press made a sensation out of this since the journalists contorted the result and presented it as an unprecedented breakthrough in practical computational methods (giving the Soviets a technological edge over the West...).[1] However, the ellipsoid method has never been interesting for the practice of linear programming—Khachyian's discovery was indeed extremely significant, but for the theory of computational complexity. It solved an open problem that many people had attacked in vain for many years. The solution was conceptually utterly different from previous approaches, which were mostly variations of the simplex method.

Input size and polynomial algorithms. In order to describe what we mean by a polynomial algorithm for linear programming, we have to define the **input size** of a linear program. Roughly speaking, it is the total number of bits needed for writing down the input to a linear programming algorithm.

First we define the *bit size* of an integer i as

$$\langle i \rangle = \lceil \log_2(|i| + 1) \rceil + 1,$$

which is the number of bits of i written in binary, including one bit for the sign. For a rational number r, i.e., a fraction $r = p/q$, the bit size is defined as $\langle r \rangle = \langle p \rangle + \langle q \rangle$. For an n-component rational vector $\mathbf{v}$ we put $\langle \mathbf{v} \rangle = \sum_{i=1}^{n} \langle v_i \rangle$,

---

[1] The following quotation from

  E. L. Lawler: The Great Mathematical Sputnik of 1979, *Math. Intelligencer* 2(1980) 191–198,

which is a remarkable article about the history of Khachyian's result, is not only of historical interest:

  The *Times* story appears to have been based on certain unshakable preconceptions of its writer, Malcolm W. Browne. Browne called George Dantzig, of Stanford University, a great pioneering authority on linear programming, and tried to force him into various admissions. Dantzig's version of the interview bears repeating:

  "What about the traveling salesman problem?" asked Browne. "If there is a connection, I don't know what it is," said Dantzig. ("The Russian discovery proposed an approach for [solving] a class of problems related to the "Traveling Salesman Problem," reported Browne.) "What about cryptography?" asked Browne. "If there is a connection, I don't know what it is," said Dantzig. ("The theory of codes could eventually be affected," reported Browne.) "Is the Russian method practical?" asked Browne. "No," said Dantzig. ("Mathematicians describe the discovery ... as a method by which computers can find solutions to a class of very hard problems that has hitherto been attacked on a hit-or-miss basis," reported Browne.)

and similarly, $\langle A \rangle = \sum_{i=1}^{m} \sum_{j=1}^{n} \langle a_{ij} \rangle$ for a rational $m \times n$ matrix A. If we consider a linear program L, say in the form

$$\text{maximize } \mathbf{c}^T \mathbf{x} \text{ subject to } A\mathbf{x} \le \mathbf{b},$$

and if we restrict ourselves to the case of A, $\mathbf{b}$, and $\mathbf{c}$ rational (which is a reasonable assumption from a computational perspective), then the bit size of L is $\langle L \rangle = \langle A \rangle + \langle \mathbf{b} \rangle + \langle \mathbf{c} \rangle$.

We say that an algorithm is a **polynomial algorithm for linear programming** if a polynomial $p(x)$ exists such that for every linear program L with rational A, $\mathbf{b}$, and $\mathbf{c}$ the algorithm finds a correct solution in at most $p(\langle L \rangle)$ steps. The steps are counted in some of the usual models of computation, for example, as steps of a Turing machine (usually the chosen computational model is not crucial; whatever is polynomial in one model is also polynomial in other reasonable models). We stress right away that a single arithmetic operation is not counted as a single step here! We count as steps operations with single bits, and hence, addition of two k-bit integers requires at least k steps.

Let us digress briefly from linear programming and let us consider Gaussian elimination, a well-known algorithm for solving systems of linear equations. For a system $A\mathbf{x} = \mathbf{b}$, where (for simplicity) A is an $n \times n$ matrix and both A and $\mathbf{b}$ are rational, we naturally define the input size as $\langle A \rangle + \langle \mathbf{b} \rangle$. Is Gaussian elimination a polynomial algorithm? This is a tricky question! Although this algorithm needs only order of n^3 arithmetic operations, the catch is that too large intermediate values could come up during the computation, even if all entries in A and in $\mathbf{b}$ are small integers. If, for example, integers with as many as 2^n bits ensued, which can indeed happen in a naive implementation of Gaussian elimination, the computation would need exponentially many steps, although it would involve only n^3 arithmetic operations. (All of this concerns *exact* computations, while many implementations use floating-point arithmetic and hence the numbers are continually rounded. But then there is no guarantee that the results are correct.) We do not want to scare the reader needlessly: It is known how Gaussian elimination can be implemented in polynomial time. We want only to point out that this is not self-evident (and not too simple, either), and call attention to one kind of trouble that may develop in attempts at proving polynomiality.

The ellipsoid method, as well as some of the interior point methods, are polynomial, while the simplex method with Bland's rule (and with many other pivot rules too) is not polynomial.[2]

---

[2] The nonpolynomiality is proved by means of the Klee–Minty cube; see Section 5.9. One has to check that an n-dimensional Klee–Minty cube can be represented by input of size polynomial in n.

Strongly polynomial algorithms. For algorithms whose input is described by a sequence of integers or rationals, such as algorithms for linear programming, the number of arithmetic operations (addition, subtraction, multiplication, division, exponentiation) is also considered, together with the number of bit operations. This often gives a more realistic picture of the running time, because contemporary computers usually execute an arithmetic operation as an elementary step, provided that the operands are not too large.

A suitable implementation of Gaussian elimination is, on the one hand, a polynomial algorithm in the sense discussed above, and on the other hand, the number of arithmetic operations is bounded by a polynomial, namely by the polynomial Cn^3 for a suitable constant C, where n is the number of equations in the system and also the number of variables. The number of arithmetic operations thus depends only on n, and it is the same for input numbers with 10 bits as for input numbers with a million bits. We say that Gaussian elimination is a **strongly polynomial** algorithm for solving systems of linear equations.

A strongly polynomial algorithm for linear programming would be one that, first, would be polynomial in the sense defined above, and second, for every linear program with n variables and m constraints it would find a solution using at most $p(m+n)$ arithmetic operations, where $p(x)$ is a fixed polynomial. But no strongly polynomial algorithm for linear programming is known, and finding one is a major open problem.

The ellipsoid method is not strongly polynomial. For every natural number M one can find a linear program with only 2 variables and 2 constraints for which the ellipsoid method executes at least M arithmetic operations (the coefficients in such linear programs must have bit size tending to infinity as $M \to \infty$). In particular, the number of arithmetic operations for the ellipsoid method cannot be bounded by any polynomial in $m+n$.

Ellipsoids. A two-dimensional ellipsoid is an ellipse plus its interior. An ellipsoid in general can most naturally be introduced as an affine transformation of a ball. We let

$$B^n = \{\mathbf{x} \in \mathbb{R}^n : \mathbf{x}^T\mathbf{x} \le 1\}$$

be the n-dimensional ball of unit radius centered at $\mathbf{0}$. Then an n-dimensional **ellipsoid** is a set of the form

$$E = \{M\mathbf{x} + \mathbf{s} : \mathbf{x} \in B^n\},$$

where M is a nonsingular $n \times n$ matrix and $\mathbf{s} \in \mathbb{R}^n$ is a vector. The mapping $\mathbf{x} \mapsto M\mathbf{x} + \mathbf{s}$ is a composition of a linear function and a translation; this is called an **affine map**.

By manipulating the definition we can describe the ellipsoid by an in-equality:

$$E = \{\mathbf{y} \in \mathbb{R}^n : M^{-1}(\mathbf{y} - \mathbf{s}) \in B^n\}$$
$$= \{\mathbf{y} \in \mathbb{R}^n : (\mathbf{y} - \mathbf{s})^T (M^{-1})^T M^{-1}(\mathbf{y} - \mathbf{s}) \leq 1\}$$
$$= \{\mathbf{y} \in \mathbb{R}^n : (\mathbf{y} - \mathbf{s})^T Q^{-1}(\mathbf{y} - \mathbf{s}) \leq 1\}, \tag{7.1}$$

where we have set $Q = MM^T$. It is well known and easy to check that such a Q is a positive definite matrix, that is, a symmetric square matrix satisfying $\mathbf{x}^T Q \mathbf{x} > 0$ for all nonzero vectors $\mathbf{x}$. Conversely, from matrix theory it is known that each positive definite matrix Q can be factored as $Q = MM^T$ for some nonsingular square matrix M. Therefore, an equivalent definition is that an ellipsoid is a set described by (7.1) for some positive definite Q and some $\mathbf{s}$.

Geometrically, $\mathbf{s}$ is the center of the ellipsoid E. If Q is a diagonal matrix and $\mathbf{s} = \mathbf{0}$, then we have an ellipsoid in *axial position*, of the form

$$\left\{ \mathbf{y} \in \mathbb{R}^n : \frac{y_1^2}{q_{11}} + \frac{y_2^2}{q_{22}} + \cdots + \frac{y_n^2}{q_{nn}} \leq 1 \right\}.$$

The axes of this ellipsoid are parallel to the coordinate axes. The numbers $\sqrt{q_{11}}, \sqrt{q_{22}}, \ldots, \sqrt{q_{nn}}$ are the lengths of the *semiaxes* of the ellipsoid E, which may sound familiar to those accustomed to the equation of an ellipse of the form $\frac{x^2}{a^2} + \frac{y^2}{b^2} = 1$. As is taught in linear algebra in connection with eigenvalues, each positive definite matrix Q can be diagonalized by an orthogonal basis change. That is, there exists an orthogonal matrix T such that the matrix TQT^{-1} is diagonal, with the eigenvalues of Q on the diagonal. Geometrically, T represents a rotation of the coordinate system that brings the ellipsoid into axial position.

A lion in the Sahara. A traditional mathematical anecdote gives directions for hunting a lion in the Sahara (under the assumption that there is at most one). We fence all of the Sahara, we divide it into two halves by another fence, and we detect one half that has no lion in it. Then we divide the other half by a fence, and we continue in this manner until the fenced piece of ground is so small that the lion cannot move and so it is caught, or if there is no lion in it, we have proved that there was none in the Sahara either. Although the qualities of this hunting guide can be disputed, for us it is essential that it gives a reasonably good description of the ellipsoid method. But in the real ellipsoid method we insist that the currently fenced piece is always an ellipsoid, even at the price that the lion can sometimes return to places from where it was expelled earlier; it is only guaranteed that the area of its territory shrinks all the time.

The ellipsoid method doesn't directly solve a linear program, but rather it seeks a solution of a system of linear inequalities $A\mathbf{x} \leq \mathbf{b}$. But as we know, this is sufficient for solving a linear program (see Section 6.1). For a simpler exposition we will first consider the following softened version of the problem:

Together with the matrix A and vector $\mathbf{b}$ we are given rational numbers $R > \varepsilon > 0$. We assume that the set $P = \{\mathbf{x} \in \mathbb{R}^n : A\mathbf{x} \leq \mathbf{b}\}$ is contained in the ball $B(\mathbf{0}, R)$ centered at $\mathbf{0}$ with radius R. If P contains a ball of radius ε, then the algorithm has to return a point $\mathbf{y} \in P$. However, if P contains no ball of radius ε, then the algorithm may return either some $\mathbf{y} \in P$, or the answer NO SOLUTION.

The ball $B(\mathbf{0}, R)$ thus plays the role of the Sahara and we assume that the lion, if present, is at least ε large. If there is only a smaller lion in the Sahara, it may escape or we may catch it—we don't care.

Under these assumptions, the ellipsoid method generates a sequence of ellipsoids $E_0, E_1, \ldots, E_t$, where $P \subseteq E_k$ for each k, as follows:

1. Set $k = 0$ and $E_0 = B(\mathbf{0}, R)$.
2. Let the current ellipsoid E_k be of the form $E_k = \{\mathbf{y} \in \mathbb{R}^n : (\mathbf{y} - \mathbf{s}_k)^T Q_k^{-1} (\mathbf{y} - \mathbf{s}_k) \leq 1\}$. If $\mathbf{s}_k$ satisfies all inequalities of the system $A\mathbf{x} \leq \mathbf{b}$, return $\mathbf{s}_k$ as a solution; **stop**.
3. Otherwise, choose an inequality of the system that is violated by $\mathbf{s}_k$. Let it be the ith inequality; so we have $\mathbf{a}_i^T \mathbf{s}_k > b_i$. Define E_{k+1} as the ellipsoid of the smallest possible volume containing the "half-ellipsoid" $H_k = E_k \cap \{\mathbf{x} \in \mathbb{R}^n : \mathbf{a}_i^T \mathbf{x} \leq \mathbf{a}_i^T \mathbf{s}_k\}$; see the following picture:

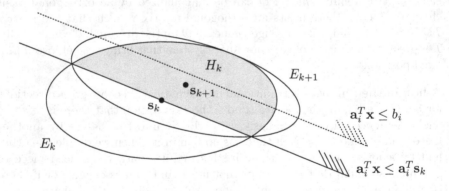

4. If the volume of E_{k+1} is smaller than the volume of a ball of radius ε, return NO SOLUTION; **stop**. Otherwise, increase k by 1 and continue with Step 2.

Let H_k' denote the intersection of the ellipsoid E_k with the half-space $\{x \in \mathbb{R}^n : \mathbf{a}_i^T \mathbf{x} \leq b_i\}$ defined by the ith inequality of the system. If $P \subseteq E_k$, then also $P \subseteq H'_k$, and the more so $P \subseteq H_k$. Why is the smallest ellipsoid containing H_k taken for E_{k+1}, instead of the smallest ellipsoid containing H'_k? Purely in order to simplify the analysis of the algorithm, since the equation of E_{k+1} comes out less complicated this way.

The ellipsoid E_{k+1}, i.e., the ellipsoid of the smallest volume containing H_k, is always determined uniquely. For illustration we mention that it is given by

$$\mathbf{s}_{k+1} = \mathbf{s}_k - \frac{1}{n+1} \cdot \frac{Q\mathbf{s}_k}{\sqrt{\mathbf{s}_k^T Q \mathbf{s}_k}},$$

$$Q_{k+1} = \frac{n^2}{n^2 - 1} \left(Q_k - \frac{2}{n+1} \cdot \frac{Q\mathbf{s}_k \mathbf{s}_k^T Q}{\mathbf{s}_k^T Q \mathbf{s}_k} \right).$$

We also leave without proof a fact crucial for the proof of correctness and efficiency of the ellipsoid method: We always have

$$\frac{\text{volume}(E_{k+1})}{\text{volume}(E_k)} \leq e^{-1/(2n+2)}.$$

Hence the volume of the ellipsoid E_k is at least $e^{k/(2n+2)}$ times smaller than the volume of the initial ball $B(\mathbf{0}, R)$. Since the volume of an n-dimensional ball is proportional to the nth power of the radius, for k satisfying $R \cdot e^{-k/n(2n+2)} < \varepsilon$ the volume of E_k is smaller than that of a ball of radius ε. Such k provides an upper bound of $\lceil n(2n+2) \ln(R/\varepsilon) \rceil$ on the maximum number of iterations. This is bounded by a polynomial in $n + \langle R \rangle + \langle \varepsilon \rangle$.

So much for the simple and beautiful idea of the ellipsoid method—now we are coming to manageable but unpleasant complications. First of all, we cannot compute the ellipsoid E_{k+1} exactly, at least not in rational arithmetic, since the defining formulas contain square roots. To get around this, E_{k+1} is computed only approximately with a suitable precision. But one has to be careful so that P is still guaranteed to be contained in E_{k+1}, and thus the approximate E_{k+1} has to be expanded slightly.

Another trouble arises, for example, when the same inequality is used for cutting the current ellipsoid in many iterations in a row. Then the ellipsoids may become too long (needle-like), and they have to be shortened artificially.

Yet another problem is that we don't really want to solve the softened problem with R and ε, but an arbitrary system of linear inequalities without any restrictions. Here the bound on the bit size of the entries of A and $\mathbf{b}$ comes into play, through the following facts:

(E1) (If a solution exists, then there is a not too large solution.) Let $\varphi = \langle A \rangle + \langle \mathbf{b} \rangle$ denote the input size for the system $A\mathbf{x} \leq \mathbf{b}$. Then this system has a solution if and only if the system

$$Ax \leq b$$
$$-K \leq x_1 \leq K$$
$$-K \leq x_2 \leq K$$
$$\vdots$$
$$-K \leq x_n \leq K$$

has a solution, where $K = 2^\varphi$. Clearly, all solutions of the latter system are contained in the ball $B(\mathbf{0}, R)$, where $R = K\sqrt{n}$.

(E2) (If a solution exists, then the solution set of a slightly relaxed system contains a small ball.) Let us put $\eta = 2^{-5\varphi}$, $\varepsilon = 2^{-6\varphi}$, and let $\boldsymbol{\eta}$ be the n-component vector with all components equal to η. Then the system $Ax \leq b$ has a solution if and only if the system $Ax \leq b + \boldsymbol{\eta}$ has a solution, and in such case the solution set of the latter system contains a ball of radius ε.

It is not hard to see how these facts can be used for solving an arbitrary system $Ax \leq b$ by the ellipsoid method. Instead of this system we solve the softened problem by the ellipsoid method, but for a new system $Ax \leq b + \boldsymbol{\eta}$, $-K - \eta \leq x_1 \leq K + \eta$, $-K - \eta \leq x_2 \leq K + \eta, \ldots, -K - \eta \leq x_n \leq K + \eta$, where K, R, ε, and η are chosen suitably (first we add the constraints $-K \leq x_i \leq K$ as in (E1), and then we apply (E2) to the resulting system). It is important that the bit size of R and ε, as well as the input size of the new system, are bounded by a polynomial function of φ. Thus the ellipsoid method runs in polynomial time, and it always finds a solution of $Ax \leq b$ if it exists.

We will not prove facts (E1) and (E2) here, but we sketch the basic ideas. For (E1) we first discuss the case $n = 2$ (in the plane). Let us consider a system of m inequalities

$$a_{i1}x + a_{i2}y \leq b_i, \quad i = 1, 2, \ldots, m.$$

Let ℓ_i be the line $\{(x, y) \in \mathbb{R}^2 : a_{i1}x + a_{i2}y = b_i\}$. It is easy to calculate that the intersection of ℓ_i and ℓ_j, if it exists, has coordinates

$$\left(\frac{a_{i2}b_j - a_{j2}b_i}{a_{i2}a_{j1} - a_{i1}a_{j2}}, \frac{a_{j1}b_i - a_{i1}b_j}{a_{i2}a_{j1} - a_{i1}a_{j2}} \right).$$

If, for example, all a_{ij} and b_i are integers with absolute value at most 1000, then the coordinates of all intersections are fractions with numerators and denominators bounded by $2 \cdot 10^6$ in absolute value. Thus, if the solution set of the considered system of inequalities has at least one vertex, such a vertex has to lie in the square $[-2 \cdot 10^6, 2 \cdot 10^6]^2$. If the solution set has no vertex and it is nonempty, it can be shown that it has to contain one of the lines ℓ_i, and that each ℓ_i intersects the just-mentioned square. Fact (E1) can be verified along these lines for the considered system with two variables. For a general system in dimension n the idea is the same, and Cramer's rule and a bound on the magnitude of the determinant of a matrix are used for estimating the coordinates of vertices of the solution set.

Fact (E2) requires more work, but the idea is similar. Each solution $\tilde{\mathbf{x}}$ of the original system $A\mathbf{x} \leq \mathbf{b}$ also satisfies the modified system $A\mathbf{x} \leq \mathbf{b} + \boldsymbol{\eta}$, and all $\mathbf{x}$ from the ball $B(\tilde{\mathbf{x}}, \varepsilon)$ satisfy it as well, because changing $\mathbf{x}$ by ε cannot change any coordinate of $A\mathbf{x}$ by more than η.

If $A\mathbf{x} \leq \mathbf{b}$ has no solution, then by a suitable variant of the Farkas lemma, namely, Proposition 6.4.3(iii), there exists a nonnegative $\mathbf{y} \in \mathbb{R}^m$ such that $\mathbf{y}^T A = \mathbf{0}^T$ and $\mathbf{y}^T \mathbf{b} < 0$, and by normalizing $\mathbf{y}$ we may assume $\mathbf{y}^T \mathbf{b} = -1$. By Cramer's rule again it is shown that there also exists a $\mathbf{y}$ with not too large components, and such $\mathbf{y}$ then witnesses unsolvability for the system $A\mathbf{x} \leq \mathbf{b} + \boldsymbol{\eta}$ as well.

Here we finish the outline of the ellipsoid method. If some parts were too incomplete and hazy for the reader, we can only recommend a more extensive treatment, for instance in the excellent book

> M. Grötschel, L. Lovász, L. Schrijver: *Geometric Algorithms and Combinatorial Optimization*, 2nd edition, Springer, Heidelberg 1994.

(We have taken the Sahara metaphor from there, among others.)

Why ellipsoids? They are used in the ellipsoid method since they constitute probably the simplest class of n-dimensional convex sets that is closed under nonsingular affine maps. Popularly speaking, this class is rich enough to approximate all convex polyhedra including flat ones and needle-like ones. If desired, ellipsoids can be replaced by simplices, for example, but the formulas in the algorithm and its analysis become considerably more unpleasant than those for ellipsoids.

The ellipsoid method need not know all of the linear program. The system of inequalities $A\mathbf{x} \leq \mathbf{b}$ can also be given by means of a **separation oracle**. This is an algorithm (black box) that accepts a point $\mathbf{s} \in \mathbb{R}^n$ as input, and if $\mathbf{s}$ is a solution of the system, it returns the answer YES, while if $\mathbf{s}$ is not a solution, it returns one (arbitrary) inequality of the system that is violated by $\mathbf{s}$. (Such an inequality separates $\mathbf{s}$ from the solution set, and hence the name separation oracle.) The ellipsoid method calls the separation oracle with the centers $\mathbf{s}_k$ of the generated ellipsoids, and it always uses the violated inequality returned by the oracle for determining the next ellipsoid.

We talk about this since a separation oracle can be implemented efficiently for some interesting optimization problems even when the full system has exponentially many inequalities or even infinitely many (so far we haven't considered infinite systems at all).

Probably the most important example of a situation in which an infinite system of linear inequalities can be solved by the ellipsoid method is **semidefinite programming**. In a semidefinite program we consider not an unknown *vector* $\mathbf{x}$, but rather an unknown square *matrix* $X = (x_{ij})_{i,j=1}^{n}$. We optimize a linear function of the variables x_{ij}. The constraints are linear inequalities and equations for the x_{ij}, *plus* the requirement that the matrix X has to be *positive semidefinite*. The last constraint distinguishes semidefinite programming from linear programming. It can be expressed by a system of infinitely many linear inequalities, namely, $\mathbf{a}^T X \mathbf{a} \geq 0$ for each $\mathbf{a} \in \mathbb{R}^n$. A separation oracle for this system can be constructed based on algorithms for computing eigenvalues and eigenvectors. The ellipsoid method can then approximate the optimum with a prescribed precision in polynomial time. (In reality, though, things are not quite as simple as it might seem from our mini-description.)

Numerous computational problems can be solved in polynomial time via semidefinite programming, some of them exactly and some approximately, and sometimes this yields the only known polynomial algorithm. A nice example of application of semidefinite programming is the **maximum cut problem** (MAXCUT), in which the vertex set of a given graph $G = (V, E)$ should be divided into two parts so that the maximum possible number of edges go between the parts. Semidefinite programming is an essential component of an approximation algorithm for MAXCUT, called the Goemans–Williamson algorithm, that always computes a partition with the number of edges going between the parts at least 87.8% of the optimal number. This is the best known approximation guarantee, and most likely also the best possible one for any polynomial algorithm. More about this and related topics can be found, for instance, in the survey

L. Lovász: Semidefinite programs and combinatorial optimization, in *Recent Advances in Algorithms and Combinatorics* (B. Reed and C. Linhares-Sales, editors), pages 137–194, Springer, New York, 2003.

Let us remark that in the just outlined applications, the ellipsoid method can be replaced by certain interior point methods (the so-called volumetric-center methods, which are not mentioned in our brief discussion of interior point methods in Section 7.2 below), and this yields algorithms efficient both in theory and in practice. See

K. Krishnan, T. Terlaky: Interior point and semidefinite approaches in combinatorial optimization, in: D. Avis, A. Hertz, and O. Marcotte (editors): *Graph Theory and Combinatorial Optimization*, Springer, Berlin etc. 2005, pages 101–158.

Theory versus practice. The notion of polynomial algorithm was suggested in the 1970s by Jack Edmonds as a formalized counterpart of the intuitive notion of an efficient algorithm. Today, a theoretician's first question for every algorithm is, Is it polynomial?

How is it possible that the ellipsoid method, which is polynomial, is much slower in practice than the simplex method, which is not polynomial? One of the reasons is that even though the ellipsoid method is polynomial, the degree of the polynomial is quite high. The second and main reason is that the simplex method is slow only on artificially constructed linear programs, which it almost never encounters in practice, while the ellipsoid method seldom behaves better than in the worst case. But the "good behavior on all inputs with rare exceptions" of the simplex method seems hard to capture theoretically. Moreover, a guaranteed efficiency for *all* inputs is much more satisfactory than only an empirically supported belief that an algorithm is usually fast.

The notion of polynomial algorithm thus has great shortcomings from a practical point of view. But attempts at constructing a polynomial algorithm in theory usually also leads, over time, to practically efficient algorithms. An impressive example in the area of linear programming are interior point methods.

7.2 Interior Point Methods

The next time linear programming made it to press headlines was in 1984. Narendra Karmakar, a researcher at IBM, suggested an algorithm that is named after him and belongs to the large family of interior point methods. He proved its polynomiality and published results of computational experiments suggesting that in practice it is much faster than the simplex method. Although his statements in the latter direction turned out to be somewhat exaggerated, interior point methods are nowadays commonly used in linear programming and often they beat the simplex method, especially on very large linear programs. They are also applied with success to semidefinite programming and other important classes of optimization problems, such as convex quadratic programming.

Interior point methods have been used for nonlinear optimization problems at least since the 1950s. For linear programs they were actually tested by the early 1970s, and interestingly, none was found competitive to the simplex method. This was because theory and hardware were not advanced enough—indeed, interior point methods typically outperform the simplex methods only on problems so large that they were beyond the capabilities of the computers at that time, and moreover, efficient implementation of interior point methods relies on powerful routines for solving large but sparse systems of linear equations, which were not available either. The success story began only with Karmakar's results.

The basic approach. When solving a linear program, the simplex method crawls along the boundary of the set of feasible solutions. The ellipsoid method encircles the set of feasible solutions, and up until the last step it remains outside of it. Interior point methods walk through the interior of the set of feasible solutions toward an optimum, carefully avoiding the boundary. Only at the very end, when they get almost to an optimum, they jump to an exact optimum by a rounding step. See the following schematic picture:

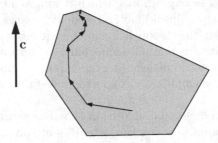

Working with an interior point all the time is the key idea that gave the methods their name. Interior points possess various pleasant mathematical properties, a kind of "nondegeneracy," and they allow one to avoid intricacies of the combinatorial structure of the boundary, which have been haunting research of the simplex method for decades. The art of interior point methods is how to keep away from the boundary while still progressing to the optimum. (For the sake of exposition, let us now assume that we are dealing with a linear program in which some initial interior point is available.)

There are several rather different basic approaches in interior point methods, and each has many variants. Interior point methods in linear programming are classified as *central path methods* (or *central trajectory methods*), *potential reduction methods*, and *affine scaling methods*, and for almost every approach one can consider a *primal version*, a *dual version*, a *primal–dual version*, or a *self-dual version*.

Here we will consider only central path methods, which have been computationally the most successful. We will present a single algorithm from this family: one with the best known theoretical complexity bounds and very good practical performance. It is fair to say, though, that a number of different interior point methods yield the same theoretical bounds and that many practitioners might say that, from their point of view, there are even better algorithms.

The analysis of the algorithm, as well as some important details of its implementation, are somewhat complicated and we will not discuss them. Modern linear programming software is based on quite advanced mathematics; for example, as was remarked above, one of the keys to the success of interior point methods is an efficient solver of special systems of linear equations, much more sophisticated than good old Gaussian elimination.

In addition to the existence of innumerable variants of interior point methods in the literature, different expositions of the same algorithm may be based on different intuition. For example, what is presented as a step of Newton's iteration method in one source may be derived by linear approximation in another source. We say this so that a reader who finds in the literature something seemingly rather different from what we say below is not confused more than necessary.

The central path. First we explain the (mathematical) notion of central path. To this end, we consider an arbitrary convex polyhedron P in $\mathbb{R}^n$ defined by a system $A\mathbf{x} \leq \mathbf{b}$ of m linear inequalities, and a linear objective function $f(\mathbf{x}) = \mathbf{c}^T\mathbf{x}$ as usual. We introduce a family of auxiliary objective functions f_μ, depending on a parameter $\mu \in [0, \infty)$:

$$f_\mu(\mathbf{x}) = \mathbf{c}^T\mathbf{x} + \mu \cdot \sum_{i=1}^{m} \ln\left(b_i - \mathbf{a}_i\mathbf{x}\right),$$

where $\mathbf{a}_i$ is the ith row of the matrix A (regarded here as a row vector). Thus $f_0 = f$ is the original linear objective function, while the f_μ for $\mu > 0$ are nonlinear, due to the logarithms. The function f_μ is constructed in such a way that when $\mathbf{x}$ approaches the boundary of P, i.e., the difference of the right-hand side and left-hand side of some inequality approaches 0, then f_μ tends to $-\infty$. The expression following μ in the definition of f_μ is called a *barrier function*, or more definitely, a *logarithmic barrier*. The word barrier is more fitting for a minimization problem, in our case minimizing $-f_\mu$, where the graph of the objective function has barriers preventing minimization algorithms from hitting the walls.

We thus consider the auxiliary problem of maximizing $f_\mu(\mathbf{x})$ over P, for given $\mu > 0$. Since f_μ is undefined on the boundary of P, we actually maximize over $\text{int}(P)$, the interior of P, and so for the problem to make sense, we need to assume that $\text{int}(P) \neq \emptyset$.

If we assume that, moreover, P is bounded, then f_μ attains a maximum at a unique point in the interior of P, which we denote by $\mathbf{x}^*(\mu)$.

Indeed, the existence of a maximum follows from the well-known fact that every continuous function attains a maximum on a compact set: The appropriate compact set is $\{\mathbf{x} \in \text{int}(P) : f_\mu(\mathbf{x}) \geq f_\mu(\mathbf{x}_0)\}$, where $\mathbf{x}_0 \in \text{int}(P)$ is arbitrary (a little argument, which we leave to the reader, is needed to verify that this set is closed).

As for uniqueness, let us assume that f_μ attains a maximum at two distinct feasible points $\mathbf{x}$ and $\mathbf{y}$. Then $f_\mu(\mathbf{x}) = f_\mu(\mathbf{y})$, and since f_μ is easily seen to be concave (meaning that $-f_\mu$ is convex), it follows that f_μ has to be constant on the segment $\mathbf{xy}$. Since the logarithm is strictly concave, this can happen only if $A\mathbf{x} = A\mathbf{y}$. But then P would contain all of the line $\{\mathbf{x} + s(\mathbf{y} - \mathbf{x}) : s \in \mathbb{R}\}$ and would not be bounded—a contradiction. In short, the maximum is unique because f_μ is strictly concave.

The condition of P bounded can be relaxed somewhat, but some condition forcing f_μ to be bounded above on P is clearly necessary, as is documented by $P = \{x \in \mathbb{R}^1 : x_1 \geq 0\}$ and either $\mathbf{c} = (1)$ or $\mathbf{c} = (0)$.

If μ is a very large number, the influence of the term $\mathbf{c}^T\mathbf{x}$ in f_μ is negligible and $\mathbf{x}^*(\mu)$ is a point "farthest from the boundary," called the *analytic center* of P. The following picture shows, for a two-dimensional P, contour lines of the function f_μ for $\mu = 100$ (the vector $\mathbf{c}$ is depicted by an arrow and the point $\mathbf{x}^*(\mu)$ by a dot):[3]

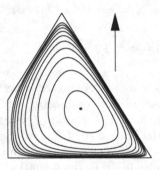

On the other hand, for small μ the point $\mathbf{x}^*(\mu)$ is close to an optimum of $\mathbf{c}^T\mathbf{x}$; see the illustrations below for $\mu = 0.5$ and $\mu = 0.1$:

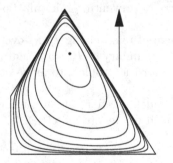 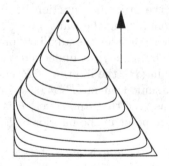

The **central path** is defined as the set $\{\mathbf{x}^*(\mu) : \mu > 0\}$. We stress that the central path is not associated with P itself, but rather with a particular system of inequalities defining P and a particular linear objective function $\mathbf{c}^T\mathbf{x}$.

---

[3] A contour line of a function $f : \mathbb{R}^2 \to \mathbb{R}$ is a set of the form $f^{-1}(\{\alpha\})$, $\alpha \in \mathbb{R}$.

The idea of central path methods is to start at $\mathbf{x}^*(\mu)$ with a suitable large μ, and then follow the central path, decreasing μ until an optimum of $\mathbf{c}^T\mathbf{x}$ is reached. Computing $\mathbf{x}^*(\mu)$ exactly would be difficult, and so actual algorithms follow the central path only approximately.

A linear program in equational form and the primal–dual central path. Having introduced the concept of central path in a geometrically transparent way, we change the setting slightly. It turns out that for the actual algorithm it is better to replace general inequality constraints by equations and nonnegativity constraints; this is similar to the case of the simplex method. So now we consider the usual linear program in equational form

$$\text{maximize } \mathbf{c}^T\mathbf{x} \text{ subject to } A\mathbf{x} = \mathbf{b}, \ \mathbf{x} \geq \mathbf{0}, \qquad (7.2)$$

where A is an $m{\times}n$ matrix of rank m. Here the barrier function should prevent violation of the nonnegativity constraints (while the equations only restrict everything to a subspace of $\mathbb{R}^n$ and they do not enter the objective function). So we set

$$f_\mu(\mathbf{x}) = \mathbf{c}^T\mathbf{x} + \mu \cdot \sum_{j=1}^{n} \ln x_j$$

and consider the auxiliary problem

$$\text{maximize } f_\mu(\mathbf{x}) \text{ subject to } A\mathbf{x} = \mathbf{b}, \ \mathbf{x} > \mathbf{0},$$

where the notation $\mathbf{x} > \mathbf{0}$ means that *all* coordinates of $\mathbf{x}$ are strictly positive.

We would again like to claim that under suitable conditions, the auxiliary problem has a unique maximizer $\mathbf{x}^*(\mu)$ for every $\mu > 0$. Obviously, we need to assume that there is a feasible $\mathbf{x} > \mathbf{0}$. Also, once we make sure that f_μ attains *at least one* maximum, it is easily seen, as above, that the maximum is unique.

We now derive *necessary* conditions for the existence of a maximum, and at the same time, we express $\mathbf{x}^*(\mu)$ as a solution of a system of equations. Later on, we will check that the necessary conditions are also sufficient. We derive these conditions by the method of **Lagrange multipliers** from analysis.

We recall that this is a general method for maximization of $f(\mathbf{x})$ subject to m constraints $g_1(\mathbf{x}) = 0, g_2(\mathbf{x}) = 0, \ldots, g_m(\mathbf{x}) = 0$, where f and $g_1, \ldots, g_m$ are functions from $\mathbb{R}^n$ to $\mathbb{R}$. It can be seen as a generalization of the basic calculus trick for maximizing a univariate function by seeking a zero of its derivative. It introduces the following system of equations with unknowns $\mathbf{x} \in \mathbb{R}^n$ and $\mathbf{y} \in \mathbb{R}^m$ (the y_i are auxiliary variables called the *Lagrange multipliers*):

$$g_1(\mathbf{x}) = g_2(\mathbf{x}) = \cdots = g_m(\mathbf{x}) = 0 \quad \text{and} \quad \nabla f(\mathbf{x}) = \sum_{i=1}^{m} y_i \nabla g_i(\mathbf{x}). \qquad (7.3)$$

Here ∇ denotes the gradient (which by convention is a *row* vector):

$$\nabla f(\mathbf{x}) = \left(\frac{\partial f(\mathbf{x})}{\partial x_1}, \frac{\partial f(\mathbf{x})}{\partial x_2}, \ldots, \frac{\partial f(\mathbf{x})}{\partial x_n} \right).$$

That is, ∇f is a vector function from $\mathbb{R}^n$ to $\mathbb{R}^n$ whose ith component is the partial derivative of f with respect to x_i. Thus, the equation $\nabla f(\mathbf{x}) = \sum_{i=1}^{m} y_i \nabla g_i(\mathbf{x})$ stipulates the equality of two n-component vectors. The method of Lagrange multipliers tells us that a maximum of $f(\mathbf{x})$ subject to $g_1(\mathbf{x}) = g_2(\mathbf{x}) = \cdots = g_m(\mathbf{x}) = 0$ occurs at $\mathbf{x}$ satisfying (7.3); that is, there exists $\mathbf{y}$ such that the considered $\mathbf{x}$ and this $\mathbf{y}$ together fulfill (7.3) (a special case of this result is derived in Section 8.7). Of course, we need some conditions on f and the g_i. It suffices to require that f and the g_i be defined on a nonempty open subset of $\mathbb{R}^n$ and have continuous first partial derivatives there, and this will be obviously satisfied in our simple application.

We apply the method of Lagrange multipliers to maximizing $f_\mu(\mathbf{x})$ subject to $A\mathbf{x} = \mathbf{b}$ (the nonnegativity constraints are taken care of implicitly, by the barriers). So we set $g_i(\mathbf{x}) = b_i - \mathbf{a}_i\mathbf{x}$. Then, after a little manipulation, the system (7.3) becomes

$$A\mathbf{x} = \mathbf{b}, \quad \mathbf{c} + \mu \left(\frac{1}{x_1}, \frac{1}{x_2}, \ldots, \frac{1}{x_n} \right) = A^T \mathbf{y}.$$

A more convenient form of this system is obtained by introducing an auxiliary nonnegative vector $\mathbf{s} = \mu \cdot \left(\frac{1}{x_1}, \frac{1}{x_2}, \ldots, \frac{1}{x_n} \right) \in \mathbb{R}^n$. We rewrite the relation of $\mathbf{s}$ and $\mathbf{x}$ to $(s_1x_1, s_2x_2, \ldots, s_nx_n) = \mu\mathbf{1}$, with $\mathbf{1}$ denoting the vector of all 1's. Then $\mathbf{x}^*(\mu)$ is expressed as the $\mathbf{x}$-part of a solution of the following system with unknowns $\mathbf{x}, \mathbf{s} \in \mathbb{R}^n$ and $\mathbf{y} \in \mathbb{R}^m$:

$$
\begin{aligned}
A\mathbf{x} &= \mathbf{b} \\
A^T\mathbf{y} - \mathbf{s} &= \mathbf{c} \\
(s_1x_1, s_2x_2, \ldots, s_nx_n) &= \mu\mathbf{1} \\
\mathbf{x}, \mathbf{s} &\geq \mathbf{0}.
\end{aligned}
\tag{7.4}
$$

All of these equations are linear except for $(s_1x_1, s_2x_2, \ldots, s_nx_n) = \mu\mathbf{1}$.

Although we have derived the system (7.4) assuming $\mu > 0$, let us make a small digression and look at what (7.4) tells us for $\mu = 0$. Then, for nonnegative $\mathbf{x}$ and $\mathbf{s}$, the equation $(s_1x_1, s_2x_2, \ldots, s_nx_n) = \mathbf{0}$ is equivalent to $\mathbf{s}^T\mathbf{x} = 0$, and since $\mathbf{s} = A^T\mathbf{y} - \mathbf{c}$, we have $0 = \mathbf{s}^T\mathbf{x} = \mathbf{y}^T A\mathbf{x} - \mathbf{c}^T\mathbf{x} = \mathbf{y}^T\mathbf{b} - \mathbf{c}^T\mathbf{x}$ (using $A\mathbf{x} = \mathbf{b}$). This may remind one of the equality of objective functions for the primal and dual linear programs. And indeed, if we take (7.2) as a primal linear program, the dual is

$$\text{minimize } \mathbf{b}^T\mathbf{y} \text{ subject to } A^T\mathbf{y} \geq \mathbf{c}, \ \mathbf{y} \in \mathbb{R}^m. \tag{7.5}$$

Then (7.4) for $\mu = 0$ tells us exactly that $\mathbf{x}$ is a feasible solution of the primal linear program (7.2), $\mathbf{y}$ is a feasible solution of the dual linear program (7.5) (here the s_j serve as slack variables expressing the difference $A^T\mathbf{y} - \mathbf{c}$), and the objective functions are equal! Hence such $\mathbf{x}$ and $\mathbf{y}$ are optimal.

So far we have shown that *if* the problem of maximizing $f_\mu(\mathbf{x})$ subject to $A\mathbf{x} = \mathbf{b}$ and $\mathbf{x} > \mathbf{0}$ has a maximum at $\mathbf{x}^*$, then there exist $\mathbf{s}^* > \mathbf{0}$ and $\mathbf{y}^* \in \mathbb{R}^m$ such that $\mathbf{x}^*, \mathbf{y}^*, \mathbf{s}^*$ satisfy (7.4). Next, we formulate conditions for the existence of the maximum (we prove only their sufficiency, but it can be shown that they are also necessary), and we show that under these conditions, the maximum is characterized by (7.4).

7.2.1 Lemma. *Let us suppose that the linear program (7.2) has a feasible solution $\tilde{\mathbf{x}} > \mathbf{0}$ and that the dual linear program (7.5) has a feasible solution $\tilde{\mathbf{y}}$ such that the slack vector $\tilde{\mathbf{s}} = A^T\tilde{\mathbf{y}} - \mathbf{c}$ satisfies $\tilde{\mathbf{s}} > \mathbf{0}$. (Less formally, both the primal and dual linear programs have an interior feasible point.) Then for every $\mu > 0$ the system (7.4) has a unique solution $\mathbf{x}^* = \mathbf{x}^*(\mu)$, $\mathbf{y}^* = \mathbf{y}^*(\mu)$, $\mathbf{s}^* = \mathbf{s}^*(\mu)$, and $\mathbf{x}^*(\mu)$ is the unique maximizer of f_μ subject to $A\mathbf{x} = \mathbf{b}$ and $\mathbf{x} > \mathbf{0}$.*

Proof. Let $\mu > 0$ be fixed. We begin with the following claim.

Claim. Under the assumptions of the lemma, the set $Q = \{\mathbf{x} \in \mathbb{R}^n : A\mathbf{x} = \mathbf{b}, \mathbf{x} > \mathbf{0}, f_\mu(\mathbf{x}) \geq f_\mu(\tilde{\mathbf{x}})\}$ is bounded.

Proof of the claim. We have

$$f_\mu(\mathbf{x}) = \mathbf{c}^T\mathbf{x} + \mu\sum_{j=1}^n \ln x_j$$
$$= \mathbf{c}^T\mathbf{x} + \tilde{\mathbf{y}}^T(\mathbf{b} - A\mathbf{x}) + \mu\sum_{j=1}^n \ln x_j \quad (\text{since } A\mathbf{x} = \mathbf{b})$$
$$= (\mathbf{c}^T - \tilde{\mathbf{y}}^T A)\mathbf{x} + \tilde{\mathbf{y}}^T\mathbf{b} + \mu\sum_{j=1}^n \ln x_j$$
$$= -\tilde{\mathbf{s}}^T\mathbf{x} + \tilde{\mathbf{y}}^T\mathbf{b} + \mu\sum_{j=1}^n \ln x_j \quad (\text{since } A^T\tilde{\mathbf{y}} - \tilde{\mathbf{s}} = \mathbf{c})$$
$$= \tilde{\mathbf{y}}^T\mathbf{b} + \sum_{j=1}^n (\mu\ln x_j - \tilde{s}_j x_j).$$

The first term of the last line is a constant, and the rest is a sum of univariate functions. Each of these univariate functions is of the form $h_\alpha(x) = \mu\ln x - \alpha x$ with $\mu, \alpha > 0$. Elementary calculus shows that $h_\alpha(x)$ attains a unique maximum at

$x = \frac{\mu}{\alpha}$, and in particular, it is bounded from above. Moreover, for every constant C, the set $\{x \in (0, \infty) : h_\alpha(x) \geq -C\}$ is bounded.

Setting $K = f_\mu(\tilde{\mathbf{x}}) - \tilde{\mathbf{y}}^T \mathbf{b}$, we have $Q \subseteq \Big\{ \mathbf{x} > \mathbf{0} :$ $\sum_{j=1}^n h_{\tilde{s}_j}(x_j) \geq K \Big\} \subseteq \prod_{j=1}^n \Big\{ x > 0 : h_{\tilde{s}_j}(x) \geq K -$ $\sum_{i \neq j} \max_{z \in \mathbb{R}} h_{\tilde{s}_i}(z) \Big\}$. The last set is a Cartesian product of bounded intervals and the claim is proved.

So the set Q is bounded, and it is not hard to check that it is closed. Hence the continuous function f_μ attains a maximum on it, which, as we know, is unique. This shows that f_μ attains a maximum under the assumptions of the lemma, and by means of Lagrange multipliers we have shown that this maximum yields a solution of (7.4). It remains to verify that this is the only solution of (7.4). What we do is to show that for every solution $\overline{\mathbf{x}}, \overline{\mathbf{y}}, \overline{\mathbf{s}}$ of (7.4), $\overline{\mathbf{x}}$ maximizes f_μ (we note that $\overline{\mathbf{s}}$ and $\overline{\mathbf{y}}$ are uniquely determined by $\overline{\mathbf{x}}$ through the relations $s_j x_j = \mu$ and $A^T \mathbf{y} - \mathbf{s} = \mathbf{c}$ from (7.4), using the assumption that A has full rank).

Let $\overline{\mathbf{x}}, \overline{\mathbf{y}}, \overline{\mathbf{s}}$ be a solution of (7.4) and let $\mathbf{x}$ satisfy $A\mathbf{x} = \mathbf{b}$ and $\mathbf{x} > \mathbf{0}$. Exactly as above we can express $f_\mu(\mathbf{x}) = \overline{\mathbf{y}}^T \mathbf{b} + \sum_{i=1}^m (\mu \ln x_j - \overline{s}_j x_j)$, and the right-hand side is maximized by setting $x_j = \mu/\overline{s}_j$, that is, for $\mathbf{x} = \overline{\mathbf{x}}$. The lemma is proved. $\square$

The set
$$\Big\{ (\mathbf{x}^*(\mu), \mathbf{y}^*(\mu), \mathbf{s}^*(\mu)) \in \mathbb{R}^{2n+m} : \mu > 0 \Big\}$$
is called the **primal–dual central path** of the linear program (7.2), and this is actually what the algorithm will follow (approximately).

The algorithm. The algorithm for solving the linear program (7.2) maintains current vectors $\mathbf{x}, \mathbf{s} \in \mathbb{R}^n$ and $\mathbf{y} \in \mathbb{R}^m$, with $\mathbf{x} > \mathbf{0}$ and $\mathbf{s} > \mathbf{0}$, that satisfy all of the *linear* equations in (7.4); that is, $A\mathbf{x} = \mathbf{b}$ and $A^T \mathbf{y} - \mathbf{s} = \mathbf{c}$. This makes sense, since it is the quadratic equations $s_j x_j = \mu$ that make the problem complicated, and moreover, these are the only equations in (7.4) in which μ enters. (We are still postponing the question of obtaining the initial $\mathbf{x}, \mathbf{y}, \mathbf{s}$.)

The current $\mathbf{x}, \mathbf{y}, \mathbf{s}$ will in general fail to satisfy the conditions $s_j x_j = \mu$, since we follow the primal–dual central path only approximately. We need to quantify by *how much* they fail to satisfy them, and one suitable "centrality" measure turns out to be
$$\mathrm{cdist}_\mu(\mathbf{x}, \mathbf{s}) = \Big\| \big(\rho(s_1 x_1, \mu), \rho(s_2 x_2, \mu), \dots, \rho(s_n x_n, \mu) \big) \Big\|,$$
where $\| \cdot \|$ is the Euclidean norm of a vector and $\rho(a, \mu) = \sqrt{a/\mu} - \sqrt{\mu/a}$. (This may look a little arbitrary, but the important thing is that it works.

Other variants of algorithms following the central path may use different distance notions.)

In a typical iteration of the algorithm, we have the current μ and $\mathbf{x}, \mathbf{y}, \mathbf{s}$ such that $\mathrm{cdist}_\mu(\mathbf{x}, \mathbf{s})$ is sufficiently small; concretely, one can take the smallness condition to be $\mathrm{cdist}_\mu(\mathbf{x}, \mathbf{s}) < \sqrt{2}$. Then we decrease μ slightly; in the considered algorithm we replace μ by $(1 - \frac{1}{2\sqrt{n}})\mu$. For this new μ we again want to approximate the solution $(\mathbf{x}^*(\mu), \mathbf{y}^*(\mu), \mathbf{s}^*(\mu))$ to (7.4) sufficiently closely. Since μ changed by only a little, we expect that the $\mathbf{x}, \mathbf{y}, \mathbf{s}$ from the previous iteration will be good initial guesses; in other words, in order to get $\mathbf{x}^*(\mu), \mathbf{y}^*(\mu), \mathbf{s}^*(\mu)$, we need to change $\mathbf{x}, \mathbf{y}, \mathbf{s}$ by only a little. Let us denote the required changes by $\Delta\mathbf{x}$, $\Delta\mathbf{y}$, and $\Delta\mathbf{s}$, respectively. So we look for $\Delta\mathbf{x}$, $\Delta\mathbf{y}$, $\Delta\mathbf{s}$ such that (7.4) is satisfied by $\mathbf{x} + \Delta\mathbf{x}$, $\mathbf{y} + \Delta\mathbf{y}$, and $\mathbf{s} + \Delta\mathbf{s}$; that is,

$$A(\mathbf{x} + \Delta\mathbf{x}) = \mathbf{b}$$
$$A^T(\mathbf{y} + \Delta\mathbf{y}) - (\mathbf{s} + \Delta\mathbf{s}) = \mathbf{c}$$
$$\big((s_1 + \Delta s_1)(x_1 + \Delta x_1), \ldots, (s_n + \Delta s_n)(x_n + \Delta x_n)\big) = \mu\mathbf{1}$$
$$\mathbf{x} + \Delta\mathbf{x} > 0, \mathbf{s} + \Delta\mathbf{s} > 0.$$

Using the fact that $\mathbf{x}, \mathbf{y}, \mathbf{s}$ satisfy the linear equations in (7.4) exactly, that is, $A\mathbf{x} = \mathbf{b}$ and $A^T\mathbf{y} - \mathbf{s} = \mathbf{c}$, the first two lines simplify to $A\Delta\mathbf{x} = \mathbf{0}$ and $A^T\Delta\mathbf{y} - \Delta\mathbf{s} = \mathbf{0}$. So far things were exact, but now we make a heuristic step: Since $\Delta\mathbf{x}$ and $\Delta\mathbf{s}$ are supposedly small compared to $\mathbf{s}$ and $\mathbf{x}$, we will neglect the second-order products $\Delta x_j \Delta s_j$ in the equation system from the third line. We will thus approximate the required changes in $\mathbf{x}$, $\mathbf{s}$, $\mathbf{y}$ by a solution to the following system:

$$A\Delta\mathbf{x} = \mathbf{0}$$
$$A^T\Delta\mathbf{y} - \Delta\mathbf{s} = \mathbf{0} \qquad (7.6)$$
$$\big(s_1\Delta x_1 + x_1\Delta s_1, \ldots, s_n\Delta x_n + x_n\Delta s_n\big) = \mu\mathbf{1} - \big(s_1 x_1, \ldots, s_n x_n\big).$$

The unknowns are $\Delta\mathbf{x}$, $\Delta\mathbf{y}$, $\Delta\mathbf{s}$, while $\mathbf{x}, \mathbf{y}, \mathbf{s}$ are regarded as constant. Hence this is a system of *linear* equations, and it is the system whose fast solution is a computational bottleneck of this interior point method. (For an actual computation the system can still be simplified by algebraic manipulations, but this is not our concern here.) In general we also need to worry about the positivity of $\mathbf{x} + \Delta\mathbf{x}$ and $\mathbf{s} + \Delta\mathbf{s}$, but in the algorithm considered below, luckily, it turns out that this is satisfied automatically—we will comment on this later.

It can be shown that passing from $\mathbf{x}$, $\mathbf{s}$, $\mathbf{y}$ to $\mathbf{x} + \Delta\mathbf{x}$, $\mathbf{y} + \Delta\mathbf{y}$, $\mathbf{s} + \Delta\mathbf{s}$, with $\Delta\mathbf{x}$, $\Delta\mathbf{y}$, $\Delta\mathbf{s}$ given by (7.6), can be regarded as a step of the Newton iterative method for solving the system (7.4).

We are ready to describe the algorithm. The input consists of a real $m \times n$ matrix A of rank m, vectors $\mathbf{b} \in \mathbb{R}^m$, $\mathbf{c} \in \mathbb{R}^n$, and a real $\varepsilon > 0$, which is a parameter controlling the accuracy of the returned solution. (If we want an exact solution, the approximate solution found by the algorithm still has to be "rounded" suitably; we will not discuss this step.)

---

1. Set $\mu = 1$ and initialize $\mathbf{x}, \mathbf{y}, \mathbf{s}$ so that $A\mathbf{x} = \mathbf{b}$, $A^T\mathbf{y} - \mathbf{s} = \mathbf{c}$, $\mathbf{x} > \mathbf{0}$, $\mathbf{s} > \mathbf{0}$, and $\text{cdist}_\mu(\mathbf{x}, \mathbf{s}) < \sqrt{2}$.
2. (Main loop) While $\mu \geq \varepsilon$, repeat Steps 3 and 4. As soon as $\mu < \varepsilon$, return $\mathbf{x}$ as an approximately optimal solution and stop.
3. Replace μ with $\left(1 - \frac{1}{2\sqrt{n}}\right)\mu$.
4. (Newton step) Compute $\Delta\mathbf{x}, \Delta\mathbf{y}, \Delta\mathbf{s}$ as the (unique) solution of the linear system (7.6). Replace $\mathbf{x}$ by $\mathbf{x} + \Delta\mathbf{x}$, $\mathbf{y}$ by $\mathbf{y} + \Delta\mathbf{y}$, and $\mathbf{s}$ by $\mathbf{s} + \Delta\mathbf{s}$. Go to the next iteration of the main loop.

Step 1 is highly nontrivial—we know that finding a feasible solution of a general linear program is computationally as hard as finding an optimal solution—and we will discuss it soon. The rest of the algorithm has been specified completely, up to the way of solving (7.6) efficiently.

What needs to be done in the analysis. Here we prove neither correctness of the algorithm nor a bound on its running time. These things are not really difficult but they are somewhat technical. We just outline what has to be done.

Let us note that the centrality measure appears only in the first step of the algorithm (initialization). In the main loop we do not explicitly check that the current $\mathbf{x}$ and $\mathbf{s}$ stay close to the central path—this has to be established in the analysis, and a similar thing holds for the conditions $\mathbf{x} > \mathbf{0}$ and $\mathbf{s} > \mathbf{0}$. In other words, one needs to show that the following *invariant* holds for the current $\mathbf{x}$ and $\mathbf{s}$ in each iteration of the main loop:

Invariant: $\mathbf{x} > \mathbf{0}$, $\mathbf{s} > \mathbf{0}$, and $\text{cdist}_\mu(\mathbf{x}, \mathbf{s}) < \sqrt{2}$.

The next item is finiteness and convergence. It turns out that the algorithm always finishes, and moreover, it needs at most $O(\sqrt{n} \log \frac{1}{\varepsilon})$ iterations. Last but not least, there is also the issue of rounding errors and numerical stability. This concludes our sketch of the analysis.

Variations. The realm of interior point methods is vast, and even the number of papers on central path methods runs into the thousands. We mention just several ideas on how the described algorithm can be varied, with the aim of better practical convergence, numerical stability, etc.

- (Higher-order methods) We have said that the computation of $\Delta\mathbf{x}, \Delta\mathbf{y}, \Delta\mathbf{s}$ is equivalent to a step of the Newton method. This method locally approximates the considered functions by linear functions, based on their first derivatives. One can also employ *higher-order methods*, where the approximation is done by multivariate polynomials, based on higher-order derivatives.

- (Truncated Newton steps) In the algorithm described above, luckily, it can be shown that making the full "Newton step," i.e., going from $\mathbf{x}, \mathbf{y}, \mathbf{s}$ to $\mathbf{x} + \Delta\mathbf{x}, \mathbf{y} + \Delta\mathbf{y}, \mathbf{s} + \Delta\mathbf{s}$, cannot leave the feasible region. For other algorithms this need not be guaranteed, and then one chooses a parameter $\alpha \in (0, 1]$ in each iteration and moves only to $\mathbf{x} + \alpha\Delta\mathbf{x}, \mathbf{y} + \alpha\Delta\mathbf{y}, \mathbf{s} + \alpha\Delta\mathbf{s}$, where α is determined so as to maintain feasibility, or also depending on other considerations.

- (Long-step methods) Decreasing μ by the factor $1 - \frac{1}{2\sqrt{n}}$ is a careful, "short-step" strategy, designed so that we do not move too far along the central path and a small change again brings $\mathbf{x}, \mathbf{y}, \mathbf{s}$ close enough to the new point of the central path. In practice it seems advantageous to make *longer steps*, i.e., to decrease μ more significantly. For example, some algorithms go from μ to $\frac{1}{10}\mu$ or so. There are even adaptive algorithms (where the new μ is not given by an explicit formula) that asymptotically achieve quadratic convergence; that is, for μ sufficiently small, a single steps goes from μ to $const \cdot \mu^2$.

  After such a large change of μ, it is in general not sufficient to make a single Newton step. Rather, one iterates Newton steps with μ fixed until the current $\mathbf{x}$ and $\mathbf{s}$ get sufficiently close to the central path.

  The theoretical analysis becomes more difficult for long-step methods, and for some of the practically most successful such algorithms no reasonable theoretical bounds are known.

Initialization. It remains to say how the first step of the algorithm, finding the initial $\mathbf{x}, \mathbf{y}, \mathbf{s}$, can be realized. There are several approaches. Here we discuss one of them, an elegant method called a *self-dual embedding*.

The idea is this: Given an input linear program, we set up another, auxiliary linear program with the following properties:

(P1) The auxiliary linear program is always feasible and bounded and there is a simple, explicitly specified vector lying on its central path, from which the above path-following algorithm can be started.

(P2) From the optimal solution of the auxiliary linear program found by the algorithm we can read off an optimal solution of the original linear program or conclude that the original linear program is infeasible or unbounded.

We develop the auxiliary linear program in several steps. Here things come out more nicely if we start with the original linear program in an inequality form:

$$\text{Maximize } \mathbf{c}^T\mathbf{x} \text{ subject to } A\mathbf{x} \leq \mathbf{b}, \ \mathbf{x} \geq \mathbf{0}. \tag{7.7}$$

As we have noted at the end of Section 6.1, the duality theorem implies that (7.7) is feasible and bounded if and only if the following system has a solution:

$$A\mathbf{x} \leq \mathbf{b}, \ A^T\mathbf{y} \geq \mathbf{c}, \ \mathbf{b}^T\mathbf{y} - \mathbf{c}^T\mathbf{x} \leq 0, \ \mathbf{x},\mathbf{y} \geq \mathbf{0}$$

(the first inequality is the feasibility of $\mathbf{x}$, the second inequality is the feasibility of $\mathbf{y}$ for the dual linear program, and the third inequality forces equality of the primal and dual objective functions). Now comes a small trick: We introduce a new scalar variable $\tau \geq 0$ (we will use Greek letters to denote such "stand-alone" scalar variables) and we multiply the right-hand sides by it. In this way, we obtain a *homogeneous* system (called the *Goldman–Tucker system*):

$$
\begin{aligned}
A\mathbf{x} - \tau\mathbf{b} &\leq \mathbf{0} \\
-A^T\mathbf{y} \qquad + \tau\mathbf{c} &\leq \mathbf{0} \\
\mathbf{b}^T\mathbf{y} - \mathbf{c}^T\mathbf{x} \qquad &\leq 0 \\
\mathbf{x},\mathbf{y} \geq \mathbf{0}, \ \tau &\geq 0.
\end{aligned}
\tag{GTS}
$$

This system, being homogeneous, always admits the zero solution, but this is uninteresting. Interesting solutions, which allow us to solve the original linear program, are those with $\tau > 0$ or $\rho > 0$, where $\rho = \rho(\mathbf{x},\mathbf{y}) = \mathbf{c}^T\mathbf{x} - \mathbf{b}^T\mathbf{y}$ denotes the slack in the last inequality. This is because of the following lemma:

7.2.2 Lemma. *No solution of (GTS) has both τ and ρ nonzero, and exactly one of the following possibilities always occurs:*

(i) *There is a solution $(\mathbf{x},\mathbf{y},\tau)$ with $\tau > 0$ (and $\rho = 0$), in which case $\frac{1}{\tau}\mathbf{x}$ is an optimal solution of the primal linear program (7.7) and $\frac{1}{\tau}\mathbf{y}$ is an optimal solution of the dual.*

(ii) *There is a solution $(\mathbf{x},\mathbf{y},\tau)$ with $\rho > 0$ (and $\tau = 0$), in which case the primal linear program (7.7) is infeasible or unbounded.*

Proof (sketch). By weak duality it is immediate that any solution with $\tau > 0$ has $\rho = 0$ and yields a pair of optimal solutions. Conversely, by the (strong) duality theorem, any pair $(\mathbf{x}^*,\mathbf{y}^*)$ of optimal solutions provides a solution of (GTS) with $\tau = 1$. Hence $\rho > 0$ implies $\tau = 0$ and infeasibility or unboundedness of the primal linear program. It remains to check that infeasibility of the primal linear program or of the dual linear program gives a solution of (GTS) with $\rho > 0$.

Let us assume, for example, that the primal linear program is infeasible (the dual case is analogous); that is, $A\mathbf{x} = \mathbf{b}$ has no nontrivial nonnegative solution. Then by the Farkas lemma (Proposition 6.4.1) there is a $\overline{\mathbf{y}}$ with $A^T\overline{\mathbf{y}} \geq \mathbf{0}$ and $\mathbf{b}^T\overline{\mathbf{y}} < 0$, and then $\mathbf{x} = \mathbf{0}$, $\mathbf{y} = \overline{\mathbf{y}}$, $\tau = 0$ is a solution to (GTS) with $\rho = -\mathbf{b}^T\overline{\mathbf{y}} > 0$.   □

As a side remark, we note that applying the Farkas lemma to the Goldman–Tucker system yields an alternative derivation of the duality theorem from the Farkas lemma.

To boost the moral, we note that we have achieved something like (P2): By means of solutions to (GTS) of a special kind we can solve the original linear program, as well as deal with its infeasibility or unboundedness. But how do we compute a solution with $\rho > 0$ or $\tau > 0$, avoiding the trivial zero solution?

Luckily, interior point methods are very suitable for this, since they converge to a "most generic" optimal solution of the given linear program. Roughly speaking, the interior point algorithm described above converges to the analytic center of the set of all optimal solutions, and this particular optimal solution does not satisfy any constraint (inequality or nonnegativity constraint) with equality if this can be avoided at all. In particular, if we made (GTS) into a linear program by adding a suitable objective function, a "most generic" optimal solution would have $\tau > 0$ or $\rho > 0$.

There is still an obstacle to be overcome: We do not have an explicit feasible interior point for (GTS) to start the algorithm, and what is worse, no feasible interior point (one with all variables and slacks strictly positive) exists! This is because we always have $\tau = 0$ or $\rho = 0$, as was noted in the lemma.

The next step is to enlarge the system so that the vector of all 1's is "forced" as a feasible interior point. Before proceeding we simplify the notation: We write the system (GTS) in matrix form as $M_0\mathbf{u} \leq \mathbf{0}$, $\mathbf{u} \geq \mathbf{0}$, where

$$M_0 = \left(\begin{array}{c|c|c} 0 & A & -\mathbf{b} \\ \hline -A^T & 0 & \mathbf{c} \\ \hline \mathbf{b} & -\mathbf{c} & 0 \end{array} \right), \quad \mathbf{u} = \left(\begin{array}{c} \mathbf{y} \\ \mathbf{x} \\ \tau \end{array} \right).$$

We note that $M_0^T = -M_0$; that is, M_0 is a *skew-symmetric matrix*. If the original matrix A has size $m \times n$, then M_0 is a $k \times k$ matrix with $k = n + m + 1$.

Now we set up the appropriate linear program with a feasible interior point, and then we gradually explain how and why it works. We define a vector $\mathbf{r} \in \mathbb{R}^k$ by $\mathbf{r} = \mathbf{1} + M_0\mathbf{1}$, a $(k+1) \times (k+1)$ matrix M by

$$M = \left(\begin{array}{c|c} M_0 & -\mathbf{r} \\ \hline \mathbf{r}^T & 0 \end{array} \right).$$

(we note that M is skew-symmetric too, which will be useful), and a vector $\mathbf{q} \in \mathbb{R}^{k+1}$ by $\mathbf{q} = (0, 0, \ldots, 0, k+1)$. We consider the following linear program with variable vector $\mathbf{v} = (\mathbf{u}, \vartheta)$ (that is, $\mathbf{u}$ with a new variable ϑ appended to the end):

$$\text{Maximize } -\mathbf{q}^T \mathbf{v} \text{ subject to } M\mathbf{v} \leq \mathbf{q}, \ \mathbf{v} \geq \mathbf{0}. \tag{SD}$$

First we explain the abbreviation (SD). It stands for *self-dual*, and indeed, if we form the dual linear program to (SD), we obtain

$$\text{minimize } \mathbf{q}^T \mathbf{w} \text{ subject to } M^T \mathbf{w} \geq -\mathbf{q}, \ \mathbf{w} \geq \mathbf{0}.$$

Using $M^T = -M$ and changing signs, which changes minimization to maximization and flips the direction of the inequality, we arrive exactly at (SD). So (SD) is equivalent to its own dual linear program.

For a feasible solution $\mathbf{v}$ of (SD) we define the *slacks* by $\mathbf{z} = \mathbf{z}(\mathbf{v}) = \mathbf{q} - M\mathbf{v}$ ("z for zlacks"). Here is a key notion: A feasible solution $\mathbf{v}$ of (SD) is called **strictly complementary** if for every $j = 1, 2, \ldots, k+1$ we have $v_j > 0$ or $z_j > 0$. The next lemma shows that a strictly complementary optimal solution is exactly what we need for solving the original linear program (7.7) (more precisely, we don't need the full force of strict complementarity, only a strict complementarity for a particular j).

7.2.3 Lemma. *The linear program (SD) is feasible and bounded, every optimal solution has $\vartheta = 0$, and hence its $\mathbf{u}$-part is a solution of the Goldman–Tucker system (GTS). Moreover, every strictly complementary optimal solution yields a solution of (GTS) with $\tau > 0$ or $\rho > 0$.*

Proof. We have $\mathbf{0}$ as a feasible solution of (SD), and also of its dual. Thus (SD) is feasible and bounded. For $\mathbf{v} = \mathbf{w} = \mathbf{0}$ both the primal and dual objective functions have value 0, so 0 must be their common optimal value. It follows that every optimal solution has $\vartheta = 0$. The rest of the lemma is easily checked. $\square$

Hence for solving the original linear program (7.7) it suffices to find a strictly complementary optimal solution of (SD). It turns out that this is exactly what the algorithm described above computes.

We will not prove this in full, but let us see what is going on. In order to apply the algorithm to (SD), we first need to convert (SD) to equational form by adding the slack variables:

$$\text{Maximize } -\mathbf{q}^T \mathbf{v} \text{ subject to } M\mathbf{v} + \mathbf{z} = \mathbf{q}, \ \mathbf{v}, \mathbf{z} \geq \mathbf{0}.$$

Now we want to write down the system (7.4) specifying points on the central path for the considered linear program. So we substitute

$A = (M \mid I_{k+1})$ (so A is a $(k+1) \times 2(k+1)$ matrix), $\mathbf{b} = \mathbf{q}$, $\mathbf{c} = (-\mathbf{q}, \mathbf{0})$, and $\mathbf{x} = (\mathbf{v}, \mathbf{z})$. We also need to introduce the extra variables $\mathbf{y}$ and $\mathbf{s}$ appearing in (7.4): For reasons that will become apparent later, we write $\mathbf{v}'$ for $\mathbf{y}$ and $(\mathbf{z}', \mathbf{z}'')$ for $\mathbf{s}$, where $\mathbf{z}'$ and $\mathbf{z}''$ are $(k+1)$-component vectors. So we have $\mathbf{v}, \mathbf{z}, \mathbf{z}', \mathbf{z}''$ nonnegative, while $\mathbf{v}'$ is (so far) arbitrary.

The equation $A\mathbf{x} = \mathbf{b}$ from (7.4) becomes $M\mathbf{v} + \mathbf{z} = \mathbf{q}$. From $A^T \mathbf{y} - \mathbf{s} = \mathbf{c}$ we get two equations: $M^T \mathbf{v}' - \mathbf{z}' = -\mathbf{q}$ and $\mathbf{v}' - \mathbf{z}'' = \mathbf{0}$. The first of these two can be rewritten as $M\mathbf{v}' + \mathbf{z}' = \mathbf{q}$. The second just means that $\mathbf{v}' = \mathbf{z}'' \geq \mathbf{0}$, and so $\mathbf{z}''$ can be disregarded if we add the constraint $\mathbf{v}' \geq \mathbf{0}$. Finally, $s_j x_j = \mu$ in (7.4) yields $v_j z_j' = \mu$ and $v_j' z_j = \mu$ for all $j = 1, 2, \ldots, k+1$. The full system is thus

$$
\begin{aligned}
M\mathbf{v} + \mathbf{z} &= \mathbf{q} \\
M\mathbf{v}' + \mathbf{z}' &= \mathbf{q} \\
v_j z_j' &= \mu \text{ for all } j = 1, 2, \ldots, k+1 \\
v_j' z_j &= \mu \text{ for all } j = 1, 2, \ldots, k+1 \\
\mathbf{v}, \mathbf{z}, \mathbf{v}', \mathbf{z}' &\geq \mathbf{0}.
\end{aligned}
$$

It is easily verified, using the skew-symmetry of M_0, that the system $M\mathbf{v} + \mathbf{z} = \mathbf{q}$ is satisfied by $\mathbf{v} = \mathbf{z} = \mathbf{1}$ (M and $\mathbf{q}$ were set up that way). Thus $\mathbf{v} = \mathbf{z} = \mathbf{v}' = \mathbf{z}' = \mathbf{1}$ is a solution of the just-derived system with $\mu = 1$, and we can use it as an initial point on the central path in Step 1 of the algorithm. To complete the analysis one needs to show that the algorithm converges to a strictly complementary optimal solution, and as we said above, this part is omitted.

As a final remark to the algorithm we note that the system above can be simplified. We know that (7.4) in general has a unique solution (provided that the primal and dual linear programs both have a feasible interior point, which is satisfied in our case). At the same time, we observe that if $\mathbf{v}, \mathbf{z}, \mathbf{v}', \mathbf{z}'$ is a solution, then interchanging $\mathbf{v}$ with $\mathbf{v}'$ and $\mathbf{z}$ with $\mathbf{z}'$ also yields a solution, and so uniqueness implies $\mathbf{v} = \mathbf{v}'$ and $\mathbf{z} = \mathbf{z}'$. Therefore, it is sufficient to work with the simpler system

$$
M\mathbf{v} + \mathbf{z} = \mathbf{q}, \; v_j z_j = \mu \text{ for } j = 1, 2, \ldots, k+1, \; \mathbf{z}, \mathbf{v} \geq \mathbf{0}
$$

and to set up the corresponding linear system for the changes $\Delta\mathbf{v}$ and $\Delta\mathbf{z}$ accordingly. This concludes the description and partial analysis of the considered interior point algorithm.

Let us remark that the self-dual embedding trick is quite universal: Almost any interior point algorithm can be used for computing an optimum of the self-dual linear program constructed as above, and this yields an optimal solution of the original linear program.

Computational complexity of interior point methods. Several interior point methods for linear programming are known to be (weakly) polynomial algorithms, including the one given above (with the self-dual embedding). The total number of bit operations for the best of these algorithms is bounded by $O(n^3 L)$, where L is the maximum bit size of coefficients in the linear program and n is the number of variables. The maximum number of iterations before reaching an optimum is $O(\sqrt{n}L)$. On the other hand, examples are known (based on the Klee–Minty cube, again!) for which any algorithm following the central path must make $\Omega(\sqrt{n}\log n)$ iterations; see

A. Deza, E. Nematollahi, and T. Terlaky: How good are interior point methods? Klee–Minty cubes tighten iteration-complexity bounds, Technical Report, McMaster University, 2004.

In practice the number of iterations seems to be bounded by a constant or by $O(\log n)$ in almost all cases. This is somewhat similar to the situation for the simplex method, where the worst-case behavior is much worse than the behavior for typical inputs.

Our presentation of interior point methods is inspired mostly by

T. Terlaky: An easy way to teach interior-point methods, *European Journal of Operational Research* 130, 1(2001), 1–19,

and full proofs of the results in this section can be found in

C. Roos, T. Terlaky, and J.-P. Vial: *Interior Point Methods for Linear Optimization (2nd edition)*, Springer, Berlin etc., 2005.

A compact survey is

F. A. Potra, S. J. Wright: Interior-point methods, *Journal of Computational and Applied Mathematics* 124(2000), pages 281–302;

at time of writing this text it was accessible at websites of the authors. Several more books from the late 1990s are cited in these papers, and an immense amount of material can be found online.

8. More Applications

Here we have collected several applications of linear programming, and in particular, of the duality theorem. They are slightly more advanced than those in Chapters 2 and 3, but we have tried to keep everything very concrete and as elementary as possible, and we hope that even a mathematically inexperienced reader will have no problems enjoying these small gems.

8.1 Zero-Sum Games

The Colonel Blotto game. Colonel Blotto and his opponent are preparing for a battle over three mountain passes. Each of them commands five regiments. The one who sends more regiments to a pass occupies it, but when the same number of regiments meet, there will be a draw. Finally, the one who occupies more passes than the other wins the battle, with a draw occurring if both occupy the same number of passes.

Given that all three passes have very similar characteristics, the strategies independently pursued by both Colonel Blotto and his opponent are the following: First they partition their five regiments into three groups. For example, the partition $(0, 1, 4)$ means that one pass will be attacked by 4 regiments, another pass by 1 regiment, and one pass will not be attacked at all. Then, the groups are assigned to the passes randomly; that is, each of the $3! = 6$ possible assignments of groups to passes is equally likely.

The partitions of Colonel Blotto and his opponent determine *winning probabilities* for both of them (in general, these do not add up to one because of possible draws). Both Colonel Blotto and his opponent want to bias the difference of these probabilities in their direction as much as possible. How should they choose their partitions?

This is an instance of a finite two-player **zero-sum game**. In such a game, each of the two players has a finite set of possible strategies (in our case, the partitions), and each pair of opposing strategies leads to a *payoff* known to both players. In our case, we define the payoff as Colonel Blotto's winning probability minus the opponent's winning probability. Whatever one of the players wins, the other player loses, and this explains the term zero-sum game. To some extent, it has become a part of common vocabulary.

When we number the strategies $1, 2, \ldots, m$ for the first player and $1, 2, \ldots, n$ for the second player, the payoffs can be recorded in the form of an $m \times n$ **payoff matrix**. In the Colonel Blotto game, the payoff matrix looks as follows, with the rows corresponding to the strategies of Colonel Blotto and the columns to the strategies of the opponent.

	$(0,0,5)$	$(0,1,4)$	$(0,2,3)$	$(1,1,3)$	$(1,2,2)$
$(0,0,5)$	0	$-\frac{1}{3}$	$-\frac{1}{3}$	-1	-1
$(0,1,4)$	$\frac{1}{3}$	0	0	$-\frac{1}{3}$	$-\frac{2}{3}$
$(0,2,3)$	$\frac{1}{3}$	0	0	0	$\frac{1}{3}$
$(1,1,3)$	1	$\frac{1}{3}$	0	0	$-\frac{1}{3}$
$(1,2,2)$	1	$\frac{2}{3}$	$-\frac{1}{3}$	$\frac{1}{3}$	0

For example, when Colonel Blotto chooses $(0, 1, 4)$ and his opponent chooses $(0, 0, 5)$, then Colonel Blotto wins (actually, without fighting) if and only if his two nonempty groups arrive at the two passes left unattended by his opponent. The probability for this to happen is $\frac{1}{3}$. With probability $\frac{2}{3}$, there will be a draw, so the difference of the winning probabilities is $\frac{1}{3} - 0 = \frac{1}{3}$.

Not knowing what the opponent is going to do, Colonel Blotto might want to choose a strategy that guarantees the highest payoff in the *worst case*. The only candidate for such a strategy is $(0, 2, 3)$: No matter what the opponent does, Colonel Blotto will get a payoff of at least 0 with this strategy, while all other strategies lead to negative payoff in the worst case. (Anticipating that a spy of the opponent might find out about his plans, he must reckon that the worst case will actually happen. The whole game is not a particularly cheerful matter anyway.) In terms of the payoff matrix, Colonel Blotto looks at the minimum in each row, and he chooses a row where this minimum is the largest possible.

Similarly, the opponent wants to choose a strategy that guarantees the lowest payoff (for Colonel Blotto) in the worst case. It turns out that $(0, 2, 3)$ is also the unique such choice for the opponent, because it guarantees that Colonel Blotto will receive payoff at most 0, while all other strategies allow him to achieve a positive payoff if he happens to guess or spy out the opponent's strategy. In terms of the payoff matrix, the opponent looks at the maximum in each column, and he chooses a column where this maximum is the smallest possible.

We note that if both Colonel Blotto and his opponent play the strategies selected as above, they both see their worst expectations come true, exactly those on which they pessimistically based their choice of strategy. Seeing the worst case happen might shatter hopes for a better outcome of the battle, but on the other hand, it is a relief. After the battle has been fought, neither Colonel Blotto nor his opponent will have to regret their choice: Even if both had known the other's strategy in advance, neither of them would have had an incentive to change his own strategy.

This is an interesting feature of this game: The strategy selected by Colonel Blotto and the strategy selected by his opponent as above are **best responses** against one another. In terms of the payoff matrix, the entry 0 in the row $((0, 2, 3)$ and column $(0, 2, 3))$ is a "saddle point"; it is a *minimum* in its row and a *maximum* in its column. A pair of strategies that are best responses against one another is called a **Nash equilibrium** of the game. As we will see next, not every game has a Nash equilibrium in this sense.

The Rock-Paper-Scissors game. Alice and Bob independently choose a hand gesture indicating either a rock, a piece of paper, or a pair of scissors. If both players choose the same gesture, the game is a draw, and otherwise, there is a cyclic pattern: Scissors beats paper (by cutting it), paper beats rock (by wrapping it up), rock beats scissors (by making it blunt). Assuming that the payoff to Alice is 1 if she wins, 0 if there is a draw, and -1 if she loses, the payoff matrix is

	rock	paper	scissors
rock	0	-1	1
paper	1	0	-1
scissors	-1	1	0

This game has no Nash equilibrium in the sense explained above. No entry of the payoff matrix is a minimum in its row and a maximum in its column at the same time. In more human terms, after every game, the player who lost may regret not to have played the gesture that would have beaten the gesture of the winner (and both may regret in the case of a draw). It is impossible for both players to fix strategies that are best responses each against the other.

But when we generalize the notion of a strategy, there *is* a way for both players to avoid regret. Both should decide *randomly*, selecting each of the gestures with probability 1/3. Even this strategy may lose, of course, but still there is no reason for regret, since with the same probability 1/3, it could have won, and the fact that it didn't is not a fault of the strategy but just bad luck. Indeed, in this way both Alice and Bob can guarantee that their payoff is 0 *in expectation*, and it is easy to see that neither of them can do better by unilaterally switching to a different behavior. We say that we have a *mixed Nash equilibrium* of the game (formally defined below).

A surprising fact is that *every* zero-sum game has a mixed Nash equilibrium. It turns out that such an equilibrium "solves" the game in the sense that it tells us (or rather, both of the two players) how to play the game optimally. As we will see in examples, the random decisions that are involved in a mixed Nash equilibrium need not give each of the possible strategies the same probability, as was the case in the very simple Rock-Paper-Scissors game. However, we will prove that suitable probability distributions always exist and, moreover, that they can be computed using linear programming.

Existence and computation of a mixed Nash equilibrium. Let us repeat the setup of zero-sum games in a more formal manner. We have two players, and we stick to calling them Alice and Bob. Alice has a set of m **pure strategies** at her disposal, while Bob has a set of n pure strategies (we assume that $m, n \geq 1$).

Then there is an $m \times n$ payoff matrix M of real numbers such that m_{ij} is Alice's gain (and Bob's loss) when Alice's ith pure strategy is played against Bob's jth pure strategy. For concreteness, we may think of Bob having to pay €m_{ij} to Alice. Of course, the situation is symmetric in that m_{ij} might be negative, in which case Alice has to pay €$-m_{ij}$ to Bob.

A **mixed strategy** of a player is a probability distribution over his or her set of pure strategies. We encode a mixed strategy of Alice by an m-dimensional vector of probabilities

$$\mathbf{x} = (x_1, \ldots, x_m), \quad \sum_{i=1}^{m} x_i = 1, \ \mathbf{x} \geq \mathbf{0},$$

and a mixed strategy of Bob by an n-dimensional vector of probabilities

$$\mathbf{y} = (y_1, \ldots, y_n), \quad \sum_{j=1}^{n} y_j = 1, \ \mathbf{y} \geq \mathbf{0}.$$

So a mixed strategy is not a particular case of a pure strategy; in the Rock-Paper-Scissors game, Alice has three possible pure strategies (rock, paper, and scissors), but infinitely many possible mixed strategies: She can choose any three nonnegative real numbers x_1, x_2, x_3 with $x_1 + x_2 + x_3 = 1$, and play rock with probability x_1, paper with probability x_2, and scissors with probability x_3. Each such triple (x_1, x_2, x_3) specifies a mixed strategy.

Given mixed strategies $\mathbf{x}$ and $\mathbf{y}$ of Alice and Bob, the *expected payoff* (expected gain of Alice) when $\mathbf{x}$ is played against $\mathbf{y}$ is

$$\sum_{i,j} m_{ij} \Pr_{\mathbf{x},\mathbf{y}}[\text{Alice plays } i, \text{Bob plays } j]$$

$$= \sum_{i,j} m_{ij} \Pr_{\mathbf{x}}[\text{Alice plays } i] \cdot \Pr_{\mathbf{y}}[\text{Bob plays } j]$$

$$= \sum_{i,j} m_{ij} x_i y_j$$

$$= \mathbf{x}^T M \mathbf{y}.$$

Now we are going to formalize the tenet of Colonel Blotto: "Prepare for the worst." When Alice considers playing some mixed strategy $\mathbf{x}$, she expects Bob to play a **best response** against $\mathbf{x}$: a strategy $\mathbf{y}$ that minimizes her expected payoff $\mathbf{x}^T M \mathbf{y}$. Similarly, for given $\mathbf{y}$, Bob expects Alice to play a strategy $\mathbf{x}$ that maximizes $\mathbf{x}^T M \mathbf{y}$.

For a fixed matrix M, these worst-case payoffs are captured by the following two functions:

$$\beta(\mathbf{x}) = \min_{\mathbf{y}} \mathbf{x}^T M \mathbf{y}, \qquad \alpha(\mathbf{y}) = \max_{\mathbf{x}} \mathbf{x}^T M \mathbf{y}.$$

So $\beta(\mathbf{x})$ is the best (smallest) expected payoff that Bob can achieve against Alice's mixed strategy $\mathbf{x}$, and similarly, $\alpha(\mathbf{y})$ is the best (largest) expected payoff that Alice can achieve against Bob's $\mathbf{y}$. It may also be worth noting that $\mathbf{y}_0$ is Bob's best response against some $\mathbf{x}$ exactly if $\mathbf{x}^T M \mathbf{y}_0 = \beta(\mathbf{x})$ (the symmetric statement for Alice is left to the reader).

Let us note that β and α are well-defined functions, since we are optimizing over compact sets. For β, say, the set of all $\mathbf{x}$ representing probability distributions is an $(m-1)$-dimensional simplex in $\mathbb{R}^m$, and hence indeed compact.

8.1.1 Definition. *A pair $(\tilde{\mathbf{x}}, \tilde{\mathbf{y}})$ of mixed strategies is a* **mixed Nash equilibrium** *of the game if $\tilde{\mathbf{x}}$ is a best response against $\tilde{\mathbf{y}}$ and $\tilde{\mathbf{y}}$ is a best response against $\tilde{\mathbf{x}}$ (the adjective "mixed" is often omitted); in formulas, this can be expressed as*

$$\beta(\tilde{\mathbf{x}}) = \tilde{\mathbf{x}}^T M \tilde{\mathbf{y}} = \alpha(\tilde{\mathbf{y}}).$$

In the Colonel Blotto game, we have even found a (pure) Nash equilibrium (a pair of pure strategies that are best responses against each other). However, the strategies themselves involved random decisions. We regard these decisions as "hard-wired" into the strategies and the payoff matrix.

Alternatively, we can consider each fixed assignment of regiments to passes as a pure strategy. Then we have a considerably larger payoff matrix, and there is no pure Nash equilibrium. Rather, we have a mixed Nash equilibrium. In practical terms it amounts to the same thing as the strategies described in the previous interpretation of the game, namely, dividing the regiments in groups of 3, 2, and 0, and sending one group to each pass at random.

These are two different views (models) of the same game, and we are free to investigate either one, although one may be more convenient or more realistic than the other.

Let us say that Alice's mixed strategy $\tilde{\mathbf{x}}$ is **worst-case optimal** if $\beta(\tilde{\mathbf{x}}) = \max_{\mathbf{x}} \beta(\mathbf{x})$. That is, Alice expects Bob to play his best response against every mixed strategy of hers, and she chooses a mixed strategy $\tilde{\mathbf{x}}$ that maximizes her expected payoff under this (pessimistic) assumption. Similarly, Bob's mixed strategy $\tilde{\mathbf{y}}$ is worst-case optimal if $\alpha(\tilde{\mathbf{y}}) = \min_{\mathbf{y}} \alpha(\mathbf{y})$.

The next simple lemma shows, among other things, that in order to attain a Nash equilibrium, both players must play worst-case optimal strategies.

8.1.2 Lemma.

(i) *We have $\max_{\mathbf{x}} \beta(\mathbf{x}) \leq \min_{\mathbf{y}} \alpha(\mathbf{y})$. Actually, for every two mixed strategies $\mathbf{x}$ and $\mathbf{y}$ we have $\beta(\mathbf{x}) \leq \mathbf{x}^T M \mathbf{y} \leq \alpha(\mathbf{y})$.*

(ii) *If the pair* $(\tilde{\mathbf{x}}, \tilde{\mathbf{y}})$ *of mixed strategies forms a mixed Nash equilibrium, then both* $\tilde{\mathbf{x}}$ *and* $\tilde{\mathbf{y}}$ *are worst-case optimal.*

(iii) *If mixed strategies* $\tilde{\mathbf{x}}$ *and* $\tilde{\mathbf{y}}$ *satisfy* $\beta(\tilde{\mathbf{x}}) = \alpha(\tilde{\mathbf{y}})$, *then they form a mixed Nash equilibrium.*

Proof. It is an amusing mental exercise to try to "see" the claims of the lemma by thinking informally about players and games. But a formal proof is routine, which is a nice demonstration of the power of mathematical formalism.

The first sentence in (i) follows from the second one, which in turn is an immediate consequence of the definitions of α and β.

In (ii), for any $\mathbf{x}$ we have $\beta(\mathbf{x}) \leq \alpha(\tilde{\mathbf{y}})$ by (i), and since $\beta(\tilde{\mathbf{x}}) = \alpha(\tilde{\mathbf{y}})$, we obtain $\beta(\mathbf{x}) \leq \beta(\tilde{\mathbf{x}})$. Thus $\tilde{\mathbf{x}}$ is worst-case optimal, and a symmetric argument shows the worst-case optimality of $\tilde{\mathbf{y}}$. This proves (ii).

As for (iii), if $\beta(\tilde{\mathbf{x}}) = \alpha(\tilde{\mathbf{y}})$, then by (i) we have $\beta(\tilde{\mathbf{x}}) = \tilde{\mathbf{x}}^T M \tilde{\mathbf{y}} = \alpha(\tilde{\mathbf{y}})$, and hence $(\tilde{\mathbf{x}}, \tilde{\mathbf{y}})$ is a mixed Nash equilibrium. The lemma is proved. $\square$

Here is the main result of this section:

8.1.3 Theorem (Minimax theorem for zero-sum games). *For every zero-sum game, worst-case optimal mixed strategies for both players exist and can be efficiently computed by linear programming. If* $\tilde{\mathbf{x}}$ *is a worst-case optimal mixed strategy of Alice and* $\tilde{\mathbf{y}}$ *is a worst-case optimal mixed strategy of Bob, then* $(\tilde{\mathbf{x}}, \tilde{\mathbf{y}})$ *is a mixed Nash equilibrium, and the number* $\beta(\tilde{\mathbf{x}}) = \tilde{\mathbf{x}}^T M \tilde{\mathbf{y}} = \alpha(\tilde{\mathbf{y}})$ *is the same for all possible worst-case optimal mixed strategies* $\tilde{\mathbf{x}}$ *and* $\tilde{\mathbf{y}}$.

The value $\tilde{\mathbf{x}}^T M \tilde{\mathbf{y}}$, the expected payoff in any Nash equilibrium, is called the **value** of the game. Together with Lemma 8.1.2(ii), we get that $(\tilde{\mathbf{x}}, \tilde{\mathbf{y}})$ forms a mixed Nash equilibrium *if and only if* both $\tilde{\mathbf{x}}$ and $\tilde{\mathbf{y}}$ are worst-case optimal.

This theorem, in a sense, tells us everything about playing zero-sum games. In particular, "Prepare for the worst" is indeed the best policy (for nontrivial reasons!). If Alice plays a worst-case optimal mixed strategy, her expected payoff is always *at least* the value of the game, no matter what strategy Bob chooses. Moreover, if Bob is well informed and plays a worst-case optimal mixed strategy, then Alice cannot secure an expected payoff *larger* than the value of the game, no matter what strategy she chooses. So there are no secrets and no psychology involved; both players can as well declare their mixed strategies in advance, and nothing changes.

Of course, if there are many rounds of the game and Alice suspects that Bob hasn't learned his lesson and doesn't play optimally, she can begin to contemplate how she could exploit this. Then psychology does come into play. However, by trying strategies that are not worst-case optimal, she is taking a risk, since she also gives Bob a chance to exploit *her*.

It remains to explain the name "minimax theorem." If we consider the equality $\beta(\tilde{\mathbf{x}}) = \alpha(\tilde{\mathbf{y}})$ and use the definitions of β, α, and of worst-case optimality of $\tilde{\mathbf{x}}$ and $\tilde{\mathbf{y}}$, we arrive at

$$\max_{\mathbf{x}} \min_{\mathbf{y}} \mathbf{x}^T M \mathbf{y} = \min_{\mathbf{y}} \max_{\mathbf{x}} \mathbf{x}^T M \mathbf{y},$$

and this is the explanation we offer.

The relation of Theorem 8.1.3 to Lemma 8.1.2(i) is similar to the relation of the duality theorem of linear programming to the weak duality theorem. And indeed, we are going to use the duality theorem in the proof of Theorem 8.1.3 in a substantial way.

Proof of Theorem 8.1.3. We first show how worst-case optimal mixed strategies $\tilde{\mathbf{x}}$ for Alice and $\tilde{\mathbf{y}}$ for Bob can be found by linear programming. Then we prove that the desired equality $\beta(\tilde{\mathbf{x}}) = \alpha(\tilde{\mathbf{y}})$ holds.

We begin by noticing that Bob's best response to a *fixed* mixed strategy $\mathbf{x}$ of Alice can be found by solving a linear program. That is, $\beta(\mathbf{x})$, with $\mathbf{x}$ a concrete vector of m numbers, is the optimal value of the following linear program in the variables $y_1, \ldots, y_n$:

$$\begin{array}{ll} \text{Minimize} & \mathbf{x}^T M \mathbf{y} \\ \text{subject to} & \sum_{j=1}^{n} y_j = 1 \\ & \mathbf{y} \geq \mathbf{0}. \end{array} \tag{8.1}$$

So we can evaluate $\beta(\mathbf{x})$. But for finding a worst-case optimal strategy of Alice we need to maximize β. Unfortunately, $\beta(\mathbf{x})$ is not a linear function, so we cannot directly formulate the maximization of $\beta(\mathbf{x})$ as a linear program. Fortunately, we can circumvent this issue by using linear programming duality.

Using the dualization recipe from Section 6.2, we write down the dual of (8.1):

$$\begin{array}{ll} \text{Maximize} & x_0 \\ \text{subject to} & M^T \mathbf{x} - \mathbf{1} x_0 \geq \mathbf{0} \end{array}$$

(this is a nice exercise in dualization). This dual linear program has only one variable x_0, since $x_1, \ldots, x_m$ are still regarded as fixed numbers. By the duality theorem, the optimal value of the dual linear program is the same as that of the primal, namely, $\beta(\mathbf{x})$.

In order to maximize $\beta(\mathbf{x})$ over all mixed strategies $\mathbf{x}$ of Alice, we set up a new linear program that optically looks exactly like the previous one, but in which $x_1, \ldots, x_m$ are regarded as variables (this works only because the constraints happen to be linear in $x_0, x_1, \ldots, x_m$):

$$\begin{aligned}
\text{Maximize} \quad & x_0 \\
\text{subject to} \quad & M^T \mathbf{x} - \mathbf{1} x_0 \geq \mathbf{0} \\
& \sum_{i=1}^{m} x_i = 1 \\
& \mathbf{x} \geq \mathbf{0}.
\end{aligned} \tag{8.2}$$

If $(\tilde{x}_0, \tilde{\mathbf{x}})$ denotes an optimal solution of this linear program, we have by construction

$$\tilde{x}_0 = \beta(\tilde{\mathbf{x}}) = \max_{\mathbf{x}} \beta(\mathbf{x}). \tag{8.3}$$

In a symmetric fashion, we can derive a linear program for solving Bob's task of computing a best strategy $\tilde{\mathbf{y}}$. We obtain the problem

$$\begin{aligned}
\text{minimize} \quad & y_0 \\
\text{subject to} \quad & M \mathbf{y} - \mathbf{1} y_0 \leq \mathbf{0} \\
& \sum_{j=1}^{n} y_j = 1 \\
& \mathbf{y} \geq \mathbf{0}
\end{aligned} \tag{8.4}$$

in the variables $y_0, y_1, \ldots, y_n$. Now an optimal solution $(\tilde{y}_0, \tilde{\mathbf{y}})$ satisfies

$$\tilde{y}_0 = \alpha(\tilde{\mathbf{y}}) = \min_{\mathbf{y}} \alpha(\mathbf{y}). \tag{8.5}$$

So both $\tilde{\mathbf{x}}$ and $\tilde{\mathbf{y}}$ are worst-case optimal strategies (and conversely, worst-case optimal strategies provide optimal solutions of the respective linear programs).

The punchline is that the two linear programs (8.2) and (8.4) are dual to each other! Again, the dualization recipe shows this. It follows that both programs have the same optimum value $\tilde{x}_0 = \tilde{y}_0$. Hence $\beta(\tilde{\mathbf{x}}) = \alpha(\tilde{\mathbf{y}})$ and $(\tilde{\mathbf{x}}, \tilde{\mathbf{y}})$ is a Nash equilibrium by Lemma 8.1.2(iii). □

Rock-Paper-Scissors revisited. To kill the time between their rare public appearances, Santa Claus and the Easter Bunny play the rock-paper-scissors game against each other. The Easter Bunny, however, cannot indicate a pair of scissors with his paw and is therefore limited to two pure strategies. The payoff matrix in this variant is

	rock	paper
rock	0	−1
paper	1	0
scissors	−1	1

We already see that Santa Claus should never play rock: For any possible gesture of the Easter Bunny, paper is a better strategy.

Let us apply the machinery we have just developed to find optimal mixed strategies for Santa Claus and the Easter Bunny. Recall that Santa Claus has to solve the linear program (8.2) to find the probability distribution $\tilde{\mathbf{x}} = (\tilde{x}_1, \tilde{x}_2, \tilde{x}_3)$ that determines his optimal strategy. At the same time, he will compute the game value $\tilde{x}_0$, his expected gain.

The linear program is

$$
\begin{aligned}
\text{maximize} \quad & x_0 \\
\text{subject to} \quad & x_2 - x_3 - x_0 \geq 0 \\
& - x_1 \quad\quad + x_3 - x_0 \geq 0 \\
& x_1 + x_2 + x_3 \quad\quad = 1 \\
& x_1, x_2, x_3 \geq 0.
\end{aligned}
$$

A (unique) optimal solution is $(\tilde{x}_0, \tilde{x}_1, \tilde{x}_2, \tilde{x}_3) = (\frac{1}{3}, 0, \frac{2}{3}, \frac{1}{3})$.
The Easter Bunny's problem (8.4) is

$$
\begin{aligned}
\text{minimize} \quad & y_0 \\
\text{subject to} \quad & - y_2 - y_0 \leq 0 \\
& y_1 \quad\quad - y_0 \leq 0 \\
& - y_1 + y_2 - y_0 \leq 0 \\
& y_1 + y_2 \quad\quad = 1 \\
& y_1, y_2 \geq 0.
\end{aligned}
$$

A (unique) optimal solution is $(\tilde{y}_0, \tilde{y}_1, \tilde{y}_2) = (\frac{1}{3}, \frac{1}{3}, \frac{2}{3})$.

Let us summarize: If both play optimally, Santa Claus wins $\frac{1}{3}$ on average (this is a scientific explanation of why Santa Claus can afford to bring more presents!). Both play paper with probability $\frac{2}{3}$. With the remaining probability $\frac{1}{3}$, Santa Claus plays scissors, while the Easter Bunny plays rock. This result is a simple but still nontrivial application of zero-sum game theory.

In retrospect, the original rock-paper-scissors game might appear rather boring, but this is relative: There is a *World RPS Society* (http://www.worldrps.com/) that holds an annual rock-paper-scissors world championship and sells a book on how to play the game.

Choosing numbers. Here is another game, which actually seems fun to play and in which the optimal mixed strategies are not at all obvious. Each of the two players independently writes down an integer between 1 and 6. Then the numbers are compared. If they are equal, the game is a draw. If the numbers differ by one, the player with the smaller number gets €2 from the one with the larger number. If the two numbers differ by two or more, the player with the larger number gets €1 from the one with the smaller number. We want to challenge the reader to compute the optimal mixed strategies for this game (by symmetry, they are the same for both players).

The Colonel Blotto game was apparently first considered by the mathematician Émile Borel in 1921 (he also served as a French minister of the navy, although only for several months). It can be considered for any number of regiments; for 6 regiments, there is still a Nash equilibrium defined by a pair of pure strategies, but for 7 regiments (this is the case stated in Borel's original paper) it becomes necessary to mix. Borel's paper does not mention the name "Colonel Blotto"; this name appears in Hubert Phillips's *Week-end Problems Book*, a collection of puzzles from 1933.

In his paper, Borel considers *symmetric* games in general. A symmetric game is defined by a payoff matrix M with $M^T = -M$. Borel erroneously states that if the number n of strategies is sufficiently large, one can construct symmetric games in which each player can secure a positive expected payoff, knowing the other player's mixed strategy. He concludes that playing zero-sum games requires psychology, on top of mathematics.

It was only in 1926 that John von Neumann, not knowing about Borel's work and its pessimistic conclusion, formally established Theorem 8.1.3.

Bimatrix games. An important generalization of finite zero-sum games is *bimatrix games*, in which both Alice and Bob want to maximize the payoff with respect to a payoff matrix of their own, A for Alice, and B for Bob (in the zero-sum case, $A = -B$). A bimatrix game also has at least one mixed Nash equilibrium: a pair of strategies $\tilde{\mathbf{x}}, \tilde{\mathbf{y}}$ that are best responses against each other, meaning that $\tilde{\mathbf{x}}^T A \tilde{\mathbf{y}} = \max_{\mathbf{x}} \mathbf{x}^T A \tilde{\mathbf{y}}$ and $\tilde{\mathbf{x}}^T B \tilde{\mathbf{y}} = \max_{\mathbf{y}} \tilde{\mathbf{x}}^T B \mathbf{y}$. We encourage the reader to find a mixed Nash equilibrium in the following variant of the modified rock-paper-scissors game played by Santa Claus and the Easter Bunny: As before, the loser pays € 1 to the winner, but in case of a draw, each player donates € 0.50 to charity.

The problem of finding a Nash equilibrium in a bimatrix game cannot be formulated as a linear program, and no polynomial-time algorithm is known for it. On the other hand, a Nash equilibrium can be computed by a variant of the simplex method, called the *Lemke–Howson* algorithm (but possibly with exponentially many pivot steps).

In general, Nash equilibria in bimatrix games are not as satisfactory as in zero-sum games, and there is no such thing as "the" game value. We know that in a Nash equilibrium, no player has an incentive to *unilaterally* switch to a different behavior. Yet, it may happen that *both* can increase their payoff by switching simultaneously, a situation that obviously cannot occur in a zero-sum game. This means that the Nash equilibrium was not optimal from the point of view of *social welfare*, and no player has a real desire of being in this particular Nash equilibrium. It may even happen that all Nash equilibria are of this suboptimal nature. Here is an example.

At each of the two department stores in town, *All the Best Deals* and *Buyer's Paradise*, the respective owner needs to decide whether to launch an advertisement campaign for the upcoming Christmas sale. If one store runs a campaign while the competitor doesn't, the expected extra revenue obtained from customers switching their preferred store (€50,000, say) and new customers (€10,000) easily outweighs the cost of the campaign (€20,000). If, on the other hand, both stores advertise, let us assume that the campaigns more or less neutralize themselves, with extra revenue coming only from new customers in an almost saturated market (€8,000 for each of the stores).

Listing the net revenues a_{ij}, b_{ij} as pairs, in units of €1,000, we obtain the following matrix, with rows corresponding to the strategies of *All the Best Deals*, and columns to *Buyer's Paradise*.

	advertise	don't advertise
advertise	$(-12, -12)$	$(40, -50)$
don't advertise	$(-50, 40)$	$(0, 0)$

If the store owners were friends, they might agree on running no campaign in order to save money (they'd better keep this agreement private in order to avoid a price-fixing charge). But if they do not communicate or mistrust each other, rational behavior will force both of them to waste money on campaigns. To see this, put yourself in the position of one of the store owners. Assuming that your competitor will not advertise, you definitely want to advertise in order to profit from the extra revenue. Assuming that the other store will advertise, you must advertise as well, in order not to lose customers. This means that you will advertise in any case. We say that the strategy "advertise" *strictly dominates* the strategy "don't advertise," so it would be irrational to play the latter.

Because the other store owner reaches the same conclusion, there will be two almost useless campaigns in the end. In fact, the pair of strategies (advertise, advertise) is the unique Nash equilibrium of the game (mixing does not help), but it is suboptimal with respect to social welfare.

In general bimatrix games, the players might not be able to reach the social optimum through rational reasoning, even if this optimum corresponds to an equilibrium of the game. This is probably the most serious deficiency of bimatrix games as models of real-world situations. An example is the *battle of the sexes*. A couple wants to go out at night. He prefers the boxing match, while she prefers the opera, but both prefer going together over going alone. If both are to decide independently where to go, there is no rational way of reaching a social optimum (namely both going out together, no matter where).

When the advertisement game is played repeatedly (after all, there is a Christmas sale every year), the situation changes. In the long run, wasting money every year is such a bad prospect that the following more cooperative behavior makes sense: In the first year, refrain from advertising, and in later years just do what the competitor did the year before. This strategy is known as TIT FOR TAT. If both stores adopt this policy, they will never waste money on campaigns; but even if one store deviates from it, there is no possibility of exploiting the competitor in the long run. It is easy to see that after a possible first loss, the one playing TIT FOR TAT can "pay" for any further loss, due to a previous loss of the competitor.

The Prisoner's Dilemma. The advertisement game is a variation of the well-known *prisoner's dilemma*, in which two convicts charged with a crime committed together are independently offered (somewhat unethical) plea bargains; if both stay silent, a lack of evidence will lead to only minor punishment. If each testifies against the other, there will be some bearable punishment. But if—and this is the unethical part— exactly one of the two testifies against the other, the betrayer will be rewarded and set free, while the one that remains silent will receive a heavy penalty. As before, rational behavior will force both convicts to testify against each other, even if they had nothing to do with the crime.

A popular introduction to the questions surrounding the prisoner's dilemma is

W. Poundstone: *Prisoner's Dilemma: John von Neumann, Game Theory, and the Puzzle of the Bomb*, Doubleday, New York 1992

(and the 1964 Stanley Kubrick movie *Dr. Strangelove or: How I Learned to Stop Worrying and Love the Bomb* might also widen one's horizons in this context).

A general introduction to game theory is

J. Bewersdorff: *Luck, Logic, and White Lies*, AK Peters, Wellesley 2004.

This book also contains the references to the work by Borel and von Neumann that we have mentioned above.

8.2 Matchings and Vertex Covers in Bipartite Graphs

Let us return to the job assignment problem from Section 3.2. There, the human resources manager of a company was confronted with the problem of filling seven positions with seven employees, where every employee has a score (reflecting qualification) for each position he or she is willing to accept. We found that the manager can use linear programming to find an assignment of employees to positions (a perfect matching) that maximizes the sum of scores. We have also promised to show how the manager can fill the positions optimally if there are *more* employees than positions.

Here we will first disregard the scores and ask under what conditions the task has any solution at all, that is, whether there is any assignment of positions to employees such that every employee gets a position she or he is willing to accept and each position is filled. We know that we can decide this by linear programming (we just invent some arbitrary scores, say all scores 100), but here we ask for a mathematical condition.

For example, let us consider the graph from Section 3.2 but slightly modified (some of the people changed their minds):

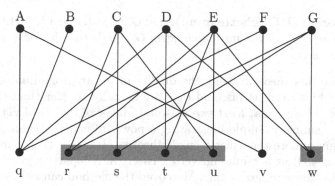

There is *no* assignment filling all positions in this situation. This is not immediately obvious, but it becomes obvious once we look at the set $\{r, s, t, u, w\}$ of jobs (marked). Indeed, the set of people willing to take *any* of these 5 jobs is $\{C, D, E, G\}$. This is only 4 people, and so they cannot be assigned to 5 different jobs.

The next theorem, known as *Hall's theorem* or the *marriage theorem*, states that if no assignment exists, we can *always* find such a simple "reason": a subset of k jobs such that the total number of employees willing to take any of them is smaller than k.

Before we formally state and prove this theorem, in the language of bipartite graphs, we need to recall the notions of maximum matching (Section 3.2) and minimum vertex cover (Section 3.3).

A matching in a graph $G = (V, E)$ is a set $E' \subseteq E$ of edges with the property that each vertex is incident to at most one edge in E'. A matching is maximum if it has the largest number of edges among all matchings in G.

A vertex cover of $G = (V, E)$ is a set $V' \subseteq V$ of vertices with the property that each edge is incident to at least one vertex in V'. A vertex cover is minimum if it has the smallest number of vertices among all vertex covers of G.

Hall's theorem gives necessary and sufficient conditions for the existence of a *best possible* matching in a bipartite graph, namely a matching that covers *all* vertices in one class of the vertex bipartition.

8.2.1 Theorem (Hall's theorem). *Let $G = (V, E)$ be a bipartite graph with bipartition $V = X \mathbin{\dot\cup} Y$. For a set $T \subseteq X$, we define the neighborhood $N(T) \subseteq Y$ of T as the set*

$$N(T) = \{w \in Y : \{v, w\} \in E \text{ for some } v \in T\}.$$

If for every $T \subseteq X$, $|N(T)| \geq |T|$ holds, then G has a matching that covers all vertices in X.

Actually, we derive the following statement, from which Hall's theorem easily follows.

8.2.2 Theorem (König's theorem). *Let $G = (V, E)$ be a bipartite graph. Then the size of a maximum matching in G equals the size of a minimum vertex cover of G.*

We prove this theorem using the duality of linear programming. There are also combinatorial proofs, and they are actually simpler than the proof offered here. There are at least two reasons in favor of the proof via duality: First, it is a simple example illustrating a powerful general technique. And second, it gives more than just König's theorem. It shows that a maximum matching, as well as a minimum vertex cover, in a bipartite graph can be computed by linear programming. Moreover, the method can be extended to computing a maximum-weight matching in a bipartite graph with weights on the edges.

Let us first see how Hall's theorem follows from König's theorem.

Proof of Theorem 8.2.1. Let $G = (X \mathbin{\dot\cup} Y, E)$ be a bipartite graph with $|N(T)| \geq |T|$ for all $T \subseteq X$. We will show that any minimum vertex cover of G has size $n_1 = |X|$. König's theorem then implies that G has a matching of size n_1, and this matching obviously covers X.

For contradiction, suppose that there is a vertex cover C with k vertices from X and fewer than $n_1 - k$ vertices from Y, for some k. The set $T = X \setminus C$ has size $n_1 - k$ and satisfies $|N(T)| \geq n_1 - k$ by the assumption. But this implies that there is a vertex $w \in N(T)$ that is not in $C \cap Y$. Since this vertex has some neighbor $v \in T$, the edge $\{v, w\}$ is not covered by C, a contradiction. □

Totally unimodular matrices. A matrix A is called **totally unimodular** if every square submatrix of A (obtained from A by deleting some rows and some columns) has determinant 0, 1, or -1. We note that, in particular, the entries of A can be only 0, -1, and $+1$.

Such matrices are interesting since an integer program with a totally unimodular constraint matrix is easy to solve—it suffices to solve its LP relaxation, as we will show in Lemma 8.2.4 below. Let us start with a preparatory step.

8.2.3 Lemma. *Let A be a totally unimodular matrix, and consider the matrix $\bar{A}$ obtained from A by appending a unit vector $\mathbf{e}_i$ as a new last column. Then $\bar{A}$ is totally unimodular as well.*

Proof. Let us fix an $\ell \times \ell$ submatrix Q of $\bar{A}$. If Q is a submatrix of A, then $\det(Q) \in \{-1, 0, 1\}$ by total unimodularity of A. Otherwise, we pick the column ℓ of Q that corresponds to the newly added column in $\bar{A}$, and we compute $\det(Q)$ according to the Laplace expansion on this column. We recall that

$$\det(Q) = \sum_{i=1}^{\ell} (-1)^{i+\ell} q_{i\ell} \det(Q^{i\ell}),$$

where $Q^{i\ell}$ is the matrix resulting from Q by removing row i and column ℓ. By construction, the ℓth column may be $\mathbf{0}$ (and we get $\det(Q) = 0$), or there is exactly one nonzero entry $q_{k\ell} = 1$. In that case,

$$\det(Q) = (-1)^{k+\ell} \det(Q^{k\ell}) \in \{-1, 0, 1\},$$

since $Q^{k\ell}$ is a submatrix of A. $\qquad\square$

The following lemma is the basis of using linear programming for solving integer programs with totally unimodular matrices.

8.2.4 Lemma. *Let us consider a linear program with n nonnegative variables and m inequalities of the form*

$$\begin{aligned} \text{maximize} \quad & \mathbf{c}^T \mathbf{x} \\ \text{subject to} \quad & A\mathbf{x} \le \mathbf{b} \\ & \mathbf{x} \ge \mathbf{0}, \end{aligned}$$

where $\mathbf{b} \in \mathbb{Z}^m$. If A is totally unimodular, and if the linear program has an optimal solution, then it also has an integral optimal solution $\mathbf{x}^* \in \mathbb{Z}^n$.

Proof. We first transform the linear program into equational form. The resulting system of equality constraints is $\bar{A}\bar{\mathbf{x}} = \mathbf{b}$, with $\bar{A} = (A \,|\, I_m)$ and $\bar{\mathbf{x}} \in \mathbb{R}^{n+m}$. Then we solve the problem using the simplex method. Let $\bar{\mathbf{x}}^*$ be an optimal basic feasible solution, associated with the feasible basis $B \subseteq \{1, 2, \dots, n+m\}$. Then we know that the nonzero entries of $\bar{\mathbf{x}}^*$ are given by

$$\bar{\mathbf{x}}_B^* = \bar{A}_B^{-1} \mathbf{b};$$

see Section 5.5.

By Cramer's rule, the entries of $\bar{A}_B^{-1}$ can be written as rational numbers with common denominator $\det(\bar{A}_B)$. The matrix $\bar{A}_B$ is a square submatrix of $\bar{A}$, where $\bar{A}$ is totally unimodular (by repeated application of Lemma 8.2.3). Since $\bar{A}_B$ is nonsingular, we get $\det(\bar{A}_B) \in \{-1, 1\}$, and the integrality of $\bar{\mathbf{x}}^*$ follows. $\qquad\square$

Incidence matrices of bipartite graphs. Here is the link between total unimodularity and König's theorem. Let $G = (X \cup Y, E)$ be a bipartite graph with n vertices $v_1, \ldots, v_n$ and m edges $e_1, \ldots, e_m$. The **incidence matrix** of G is the matrix $A \in \mathbb{R}^{n \times m}$ defined by

$$a_{ij} = \begin{cases} 1 & \text{if } v_i \in e_j \\ 0 & \text{otherwise.} \end{cases}$$

8.2.5 Lemma. Let $G = (X \cup Y, E)$ be a bipartite graph. The incidence matrix A of G is totally unimodular.

Proof. We need to prove that every $\ell \times \ell$ submatrix Q of A has determinant 0 or ± 1, and we proceed by induction on ℓ. The case $\ell = 1$ is immediate, since the entries of an incidence matrix are only 0's and 1's.

Now we consider $\ell > 1$ and an $\ell \times \ell$ submatrix Q. Since the columns of Q correspond to edges, each column of Q has at most two nonzero entries (which are 1). If there is a column with only zero entries, we get $\det(Q) = 0$, and if there is a column with only one nonzero entry, we can expand the determinant on this column (as in the proof of Lemma 8.2.3) and get that up to sign, $\det(Q)$ equals the determinant of an $(\ell-1) \times (\ell-1)$ submatrix Q'. By induction, $\det(Q') \in \{-1, 0, 1\}$, so the same holds for Q.

Finally, if every column of Q contains precisely two 1's, we claim that $\det(Q) = 0$. To see this, we observe that the sum of all rows of Q corresponding to vertices in X is the row vector $(1, \ldots, 1)$, since for each column of Q, exactly one of its two 1's comes from a vertex in X. For the same reason, we get $(1, \ldots, 1)$ by summing up the rows for vertices in Y, and it follows that the rows of Q are linearly dependent. $\qquad\square$

Now we are ready to prove König's theorem.

Proof of Theorem 8.2.2. We first consider the integer program

$$\begin{aligned} \text{maximize} \quad & \sum_{j=1}^m x_j \\ \text{subject to} \quad & A\mathbf{x} \leq \mathbf{1} \\ & \mathbf{x} \geq 0 \\ & \mathbf{x} \in \mathbb{Z}^m, \end{aligned}$$

where A is the incidence matrix of G. In this integer program, the row of A corresponding to vertex v_i induces the constraint

$$\sum_{j: e_j \ni v_i} x_j \leq 1.$$

This implies that $x_j \in \{0, 1\}$ for all j, and that the edges e_j with $\tilde{x}_j = 1$ in an optimal solution $\tilde{\mathbf{x}}$ form a maximum matching in G.

Next we consider the integer program

$$\text{minimize} \quad \sum_{i=1}^{n} y_i$$
$$\text{subject to} \quad A^T \mathbf{y} \geq \mathbf{1}$$
$$\mathbf{y} \geq \mathbf{0}$$
$$\mathbf{y} \in \mathbb{Z}^n,$$

where A is as before the incidence matrix of G. In this integer program, the row of A^T corresponding to edge e_j induces the constraint

$$\sum_{i:v_i \in e_j} y_i \geq 1.$$

This implies that in any *optimal* solution $\tilde{\mathbf{y}}$ we have $\tilde{y}_i \in \{0, 1\}$ for all i, since any larger value could be decreased to 1. But then the vertices v_i with $\tilde{y}_i = 1$ in an optimal solution $\tilde{\mathbf{y}}$ form a minimum vertex cover of G.

To summarize, the optimum value of the first integer program is the size of a maximum matching in G, and the optimum value of the second integer program is the size of a minimum vertex cover in G.

In both integer programs, we may now drop the integrality constraints without affecting the optimum values: A, and therefore also A^T, are totally unimodular by Lemma 8.2.5, and so Lemma 8.2.4 applies. But the resulting *linear* programs are dual to each other; the duality theorem thus shows that their optimal values are equal, and this proves Theorem 8.2.2.   □

It remains to explain the algorithmic implications of the proof (namely, how a maximum matching and a minimum vertex cover can actually be computed). To get a maximum matching, we simply need to find an *integral* optimal solution of the first linear program. When we use the simplex method to solve the linear program, we get this for free; see the proof of Lemma 8.2.4 and, in particular, the claim toward the end of its proof. Otherwise, we can apply Theorem 4.2.3(ii) to construct a basic feasible solution from any given optimal solution, and this basic feasible solution will be integral. A minimum vertex cover is obtained from the second (dual) linear program in the same fashion.

The previous arguments show more: Given edge weights $w_1, \ldots, w_m$, any optimal solution of the integer program

$$\text{maximize} \quad \sum_{j=1}^{m} w_j x_j$$
$$\text{subject to} \quad A\mathbf{x} \leq \mathbf{1}$$
$$\mathbf{x} \geq \mathbf{0}$$
$$\mathbf{x} \in \mathbb{Z}^m$$

corresponds to a *maximum-weight matching* in G. Since we can, as before, relax the integrality constraints without affecting the optimum value, an integral optimal solution of the relaxation can be found, and it yields a maximum-weight matching in G. This solves the optimal job assignment problem if there are more employees than jobs.

The fact that a linear program with totally unimodular constraint matrix and integral right-hand side has an integral optimal solution implies something much stronger: Since every vertex of the feasible region is optimal for some objective function (see Section 4.4), we know that *all* vertices of the feasible region are integral. We say that the feasible region forms an **integral polyhedron**.

Such integrality results together with linear programming duality can yield interesting and powerful *minimax theorems*. König's theorem is one such example. Another classical minimax theorem that can be proved along these lines is the *max-flow-min-cut theorem*.

To state this theorem, we consider a network modeled by a directed graph $G = (V, E)$ with edge capacities w_e. In Section 2.2, we have interpreted them as maximum transfer rates of data links. Given two designated vertices, the source s and the sink t, the *maximum flow* value is the maximum rate at which data can flow from s to t through the network.

The *minimum cut* value, on the other hand, is the minimum total capacity of any set of data links whose breakdown disconnects t from s.

The max-flow-min-cut theorem states that the maximum flow value is *equal* to the minimum cut value. One of several proofs writes both values as optimal values of linear programs that are dual to each other.

When we consider matchings and vertex covers in general (not necessarily bipartite) graphs, the situation changes: Total unimodularity no longer applies, and the "duality" between the two concepts disappears.

In fact, the problem of finding a minimum vertex cover in a general graph is computationally difficult (NP-hard); see Section 3.3. A maximum-weight matching, on the other hand, can still be computed in polynomial time for general graphs, although this result is by no means trivial. Behind the scenes, there is again an integrality result, based on the notion of *total dual integrality*; see the glossary.

8.3 Machine Scheduling

In the back office of the copy shop *Copy & Paste*, the operator is confronted with n copy jobs submitted by customers the night before. For processing them, she has m photocopying machines with different features at her disposal. For all i, j, the operator quickly estimates how long it would take the ith machine to process the jth job, and she makes a table of the resulting running times, like this:

	Single B&W	Duplex B&W	Duplex Color
Master's thesis, 90 pages two-sided, 10 B&W copies	—	45 min	60 min
All the Best Deals flyer, 1 page one-sided, 10,000 B&W copies	2h 45 min	4h 10 min	5h 30 min
Buyer's Paradise flyer, 1 page one-sided, 10,000 B&W copies	2h 45 min	4h 10 min	5h 30 min
Obituary, 2 pages two-sided, 100 B&W copies	—	2 min	3 min
Party platform, 10 pages two-sided, 5,000 color copies	—	—	3h 30 min

Since the operator can go home as soon as all jobs have been processed, her goal is to find an assignment of jobs to machines (a *schedule*) such that the **makespan**—the time needed to finish all jobs—is minimized. In our example, this is not hard to figure out: For the party platform, there is no choice between machines. We can also observe that it is advantageous to use both B&W machines for processing the two flyers, no matter where the thesis and the obituary go. Given this, the makespan is at least 4h 55 min if the thesis is processed on the Duplex B&W machine, so it is better put on the color machine to achieve the optimum makespan of 4h 30 min (with the obituary running on the B&W machine).

In general, finding the optimum makespan is computationally difficult (NP-hard). The obvious approach of trying all possible schedules is of course not a solution for a larger number n of jobs. What we show in this section is that the operator can quickly compute a schedule whose makespan is at most *twice* as long as the optimum makespan. All she needs for that are some linear programming skills. (To really appreciate this result, one shouldn't think of a problem with 5 jobs but with thousands of jobs.)

We should emphasize that in this scheduling problem the jobs are considered *indivisible*, and so each job must be processed on a single machine. This, in a sense, is what makes the problem difficult. As we will soon see, an optimal "fractional schedule," where a single job could be divided among several machines in arbitrary ratios, can be found efficiently by linear programming.

Two integer programs for the scheduling problem. Let us identify the m machines with the set $M := \{1, \ldots, m\}$ and the n jobs with the set $J := \{m+1, \ldots, m+n\}$.

Let d_{ij} denote the running time of job $j \in J$ on machine $i \in M$. We assume $d_{ij} > 0$. To simplify notation, we also assume that any machine can process any job: An infeasible assignment of job j to machine i can be modeled by a large running time $d_{ij} = K$. If K is larger than the sum of all "real" running times, the optimal schedule will avoid infeasible assignments, given that there is a feasible schedule at all.

With these notions, the following integer program in the variables t and x_{ij}, $i \in M$, $j \in J$ computes an assignment of jobs to machines that minimizes the makespan:

$$
\begin{aligned}
\text{Minimize} \quad & t \\
\text{subject to} \quad & \textstyle\sum_{i \in M} x_{ij} = 1 && \text{for all } j \in J \\
& \textstyle\sum_{j \in J} d_{ij} x_{ij} \leq t && \text{for all } i \in M \\
& x_{ij} \geq 0 && \text{for all } i \in M, j \in J \\
& x_{ij} \in \mathbb{Z} && \text{for all } i \in M, j \in J.
\end{aligned}
$$

Under the integrality constraints, the conditions $\sum_{i \in M} x_{ij} = 1$ and $x_{ij} \geq 0$ imply that $x_{ij} \in \{0, 1\}$ for all i, j. With the interpretation that $x_{ij} = 1$ if job j is assigned to machine i, and $x_{ij} = 0$ otherwise, the first n equations stipulate that each job is assigned to exactly one machine. The next m inequalities make sure that no machine needs more time than t to finish all jobs assigned to it. Minimizing t leads to equality for at least one of the machines, so the best t is indeed the makespan of an optimal schedule.

As we have already seen in Section 3.3 (the minimum vertex cover problem), solving the *LP relaxation* obtained from an integer program by deleting the integrality constraints can be a very useful step toward an approximate solution of the original problem. In our case, this approach needs an additional twist: We will relax another integer program, obtained by adding *redundant constraints* to the program above. After dropping integrality, the added constraints are no longer redundant and lead to a better LP relaxation.

Let t_{opt} be the makespan of an optimal schedule. If $d_{ij} > t_{\text{opt}}$, then we know that job j cannot run on machine i in any optimal schedule, so we may add the constraint $x_{ij} = 0$ to the integer program without affecting its validity. More generally, if T is any upper bound on t_{opt}, we can write down the following integer program, which has the same optimal solutions as the original one:

$$
\begin{aligned}
\text{Minimize} \quad & t \\
\text{subject to} \quad & \textstyle\sum_{i \in M} x_{ij} = 1 && \text{for all } j \in J \\
& \textstyle\sum_{j \in J} d_{ij} x_{ij} \leq t && \text{for all } i \in M \\
& x_{ij} \geq 0 && \text{for all } i \in M, j \in J \\
& x_{ij} = 0 && \text{for all } i \in M, j \in J \text{ with } d_{ij} > T \\
& x_{ij} \in \mathbb{Z} && \text{for all } i \in M, j \in J.
\end{aligned}
$$

But we do not know t_{opt}, so what is the value of T we use for the relaxation? There is no need to specify this right here; for the time being, you can imagine that we set $T = \max_{ij} d_{ij}$, a value that will definitely work, because it makes our second integer program coincide with the first one.

A good schedule from the LP relaxation. As already indicated, the first step is to solve the LP relaxation, denoted by $\text{LPR}(T)$, of our second integer program:

Minimize $\qquad\qquad t$

subject to $\quad \sum_{i \in M} x_{ij} = 1 \qquad$ for all $j \in J$

$\qquad\qquad \sum_{j \in J} d_{ij} x_{ij} \leq t \qquad$ for all $i \in M$

$\qquad\qquad\quad x_{ij} \geq 0 \qquad$ for all $i \in M, j \in J$

$\qquad\qquad\quad x_{ij} = 0 \qquad$ for all $i \in M, j \in J$ with $d_{ij} > T$.

In contrast to the vertex cover application, we cannot work with any optimal solution of the relaxation, though: We need a *basic feasible* optimal solution; see Section 4.2. To be more precise, we rely on the following property of a basic feasible solution; see Theorem 4.2.3.

8.3.1 Assumption. *The columns of the constraint matrix A corresponding to the nonzero variables x_{ij}^* in the optimal solution of* LPR(T) *are linearly independent.*

As usual, the nonnegativity constraints $x_{ij} \geq 0$ do not show up in A. In case the simplex method is used to solve the relaxation, such a solution comes for free (the column of A corresponding to t could then actually be added to the set of columns in the assumption, but this is not needed). Otherwise, we can easily construct a solution satisfying the assumption from any given optimal solution, according to the recipe in the proof of Theorem 4.2.3.

At this point the reader may wonder why we would want to use an algorithm different from the simplex method here, in particular when we are searching for a basic feasible optimal solution. The reason is of theoretical nature: We want to prove that a schedule whose makespan is at most twice the optimum can be found *in polynomial time*. As we have pointed out in the introductory part of Chapter 7, the simplex method is not known to run in polynomial time for any pivot rule, and for most pivot rules it simply does not run in polynomial time. For the theoretical result we want, we had therefore better use one of the polynomial-time methods for solving linear programs, sketched in Chapter 7. For practical purposes, the simplex method will do, of course.

In general terms, what we are trying to develop here is a poly-nomial-time *approximation algorithm* for an NP-hard problem. Since complexity theory indicates that we will not be able to solve the problem exactly within reasonable time bounds, it is quite natural to ask for an approximate solution that can be obtained in polynomial time. The quality of an approximate solution is typically measured by the *approximation factor*, the ratio between the value of the approximate solution and the value of an optimal solution. In our approximation algorithm for the scheduling problem, this factor will be at most 2.

Let us fix the values t^* and x_{ij}^* of the variables in some optimal solution of the LP relaxation. We now consider the bipartite graph $G = (M \cup J, E)$, with

$$E = \{\{i, j\} \subseteq M \cup J \mid x_{ij}^* > 0\}.$$

For an arbitrary optimal solution, this graph could easily be a (boring) complete bipartite graph, but under Assumption 8.3.1 it becomes more interesting.

8.3.2 Lemma. *In any subgraph of G (obtained by deleting edges, or vertices with their incident edges), the number of edges is at most the number of vertices.*

The graph G for $m = 4$ machines and $n = 6$ jobs might look like this, for example:

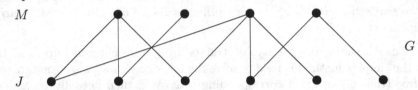

Proof. Let A be the constraint matrix of the LP relaxation. It has one row for each machine, one for each job, and one for each runtime d_{ij} exceeding T. The columns of A corresponding to the nonzero variables x_{ij}^* (equivalently, to the edges of G) are linearly independent by our assumption.

Now we consider any subgraph of G, with vertex set $M' \cup J' \subseteq M \cup J$ and edge set $E' \subseteq E$. Let A' be the submatrix obtained from A by restricting to rows corresponding to $M' \cup J'$, and to columns corresponding to E'.

Claim. *The columns of A' are linearly independent.*

> To see this, we first observe that the columns of A corresponding to E' are linearly independent, simply because $E' \subseteq E$. Any variable $x_{ij}, \{i, j\} \in E'$, occurs in the inequality for machine $i \in M'$ and in the equation for job $j \in J'$, but in no other equation, since $x_{ij}^* > 0$ implies $d_{ij} \leq T$. This means that the columns of A corresponding to E' have zero entries in all rows except for those corresponding to machines or jobs in $M' \cup J'$. Hence, these columns remain linearly independent even when we restrict them to the rows corresponding to $M' \cup J'$.

By the claim, we have $|E'| \leq |M' \cup J'|$, and this is the statement of the lemma. $\square$

This structural result about G allows us to find a good schedule.

8.3.3 Lemma. *Let $T \geq 0$ be such that the LP relaxation $\mathrm{LPR}(T)$ is feasible, and suppose that a feasible solution is given that satisfies Assumption 8.3.1 and has value $t = t^*$. Then we can efficiently construct a schedule of makespan at most $t^* + T$.*

Proof. We need to assign each job to some machine. We begin with the jobs j that have degree one in the graph G, and we assign each such j to its unique neighbor i. By the construction of G and the equation $\sum_{i \in M} x_{ij} = 1$, we have $x_{ij}^* = 1$ in this case. If machine i has been assigned a set S_i of jobs in this way, it can process these jobs in time

$$\sum_{j \in S_i} d_{ij} = \sum_{j \in S_i} d_{ij} x_{ij}^* \leq \sum_{j \in J} d_{ij} x_{ij}^* \leq t^*.$$

So each machine can handle the jobs assigned by this partial schedule in time t^*.

Next we remove all assigned jobs and their incident edges from G. This leaves us with a subgraph $G' = (M \cup J', E')$. In G', all vertices of degree one are machines. In the example depicted above, two jobs have degree one, and their deletion results in the following subgraph G':

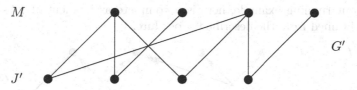

We will show that we can find a matching in G' that covers all the remaining jobs. If we assign the jobs according to this matching, every machine gets at most one additional job, and this job can be processed in time at most T by the construction of our second integer program. Therefore, the resulting full schedule has makespan at most $t^* + T$.

It remains to construct the matching. To this end, we use *Hall's theorem* from Section 8.2. According to this theorem, a matching exists if for every subset $J'' \subseteq J'$ of jobs, its neighborhood (the set of all machines connected to at least one job in J'') has size at least $|J''|$.

To check this condition, we let $J'' \subseteq J'$ be such a subset of jobs and $N(J'')$ its neighborhood. If e is the number of edges in the subgraph of G' induced by $J'' \cup N(J'')$, then Lemma 8.3.2 guarantees that $e \leq |J'' \cup N(J'')|$. On the other hand, since every job has at least two neighbors, we have $e \geq 2|J''|$, and this shows that $|N(J'')| \geq |J''|$.

Although this proof is nonconstructive, we can easily find the matching (once we know that it exists) by linear programming as in Section 8.2, or by other known polynomial-time methods. □

There is a direct way of constructing the matching in the proof of Lemma 8.3.3 that relies neither on Halls's theorem nor on general (bipartite) matching algorithms. It goes as follows: Lemma 8.3.2 implies that each connected component of G' is either a tree, or a tree with one extra edge connecting two of its vertices. In the latter case, the component has exactly one cycle of even length, because G' is bipartite. Therefore, we can match all jobs occurring on cycles, and after removing the vertices of all cycles, we are left with a subgraph G'', all of whose connected components are trees, *with at most one vertex per tree being a former neighbor of a cycle vertex*. It follows that in every tree of G'', at most one vertex can be a job of degree one, since all other degree-one vertices already had degree one in G' and are therefore machines.

The matching of the remaining jobs is easy. We root any tree in G'' at its unique job of degree one (or at any vertex if there is no such job), and we match every job to one of its children in the rooted tree. For this, we observe that there cannot be an isolated job in G'': Since a job in G'' was on no cycle in G', the removal of cycles can affect only one of the at least two neighbors of the job in G'.

For our running example, here is a complete assignment of jobs to machines obtained from the described procedure:

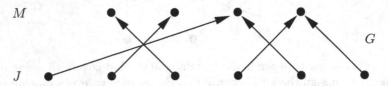

Choosing the parameter T. How good is the bound we get from the previous lemma? We will assume that t^* is the value of an optimal basic feasible solution of the LP relaxation with parameter T. Then t^* is a lower bound for the optimum makespan t_{opt}, so this part of the bound looks promising. But when we recall that T must be an *upper* bound for t_{opt} in order for our second integer program to be valid, it seems that we would have to choose $T = t_{\text{opt}}$ to get makespan at most $2t_{\text{opt}}$. But this cannot be done, since we have argued above that it is hard to compute t_{opt} (and if it were easy, there would be no need for an approximate solution anyway).

Luckily, there is a way out. Reading Lemma 8.3.3 carefully, we see that T only needs to be chosen so that the *LP relaxation* LPR(T) is feasible, and there is no need for the second *integer* program to be feasible.

If LPR(T) is feasible, Lemma 8.3.3 allows us to construct a schedule with makespan at most $t^* + T$, so the best T is the one that minimizes $t^* + T$ subject to LPR(T) being feasible. Since t^* depends on T, we make this explicit now by writing $t^* = t^*(T)$. If LPR(T) is infeasible, we set $t^*(T) = \infty$.

How to find the best T. We seek a point T^* in which the function $f(T) = t^*(T) + T$ attains a minimum.

First we observe that $t^*(T)$ is a step function as in the following picture:

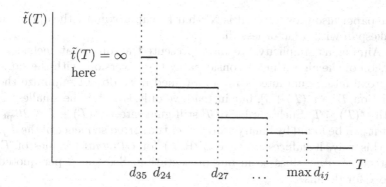

Indeed, let us start with the value $T = \max_{ij} d_{ij}$, and let us decrease T continuously. The value of $t^*(T)$ may change (jump up) only immediately after moments when a new constraint of the form $x_{ij} = 0$ appears in LPR(T), and this happens only for $T = d_{ij}$. Between these values the function $t^*(T)$ stays constant.

Consequently, the function $f(T) = t^*(T) + T$ is linearly increasing on each interval between two consecutive d_{ij}'s, and the minimum is attained at some d_{ij}. So we can compute the minimum, and the desired best value T^*, by solving at most mn linear programs of the form LPR(T), with T ranging over all d_{ij}. Under our convention that $t^*(T) = \infty$ if LPR(T) is infeasible, the minimum will be attained at a value T^* with LPR(T^*) feasible.

8.3.4 Theorem. *Let T^* be the value of T that minimizes $t^*(T) + T$. With $T = T^*$, the algorithm in the proof of Lemma 8.3.3 computes a schedule of makespan at most $2t_{\mathrm{opt}}$.*

Proof. We know that for $T = t_{\mathrm{opt}}$, the second integer program is feasible and has optimum value t_{opt}. Hence LPR(t_{opt}) is feasible as well and its optimum value can be only smaller: $t^*(t_{\mathrm{opt}}) \leq t_{\mathrm{opt}}$. We thus have

$$
\begin{aligned}
t^*(T^*) + T^* &= \min_T \left(t^*(T) + T \right) \\
&\leq t^*(t_{\mathrm{opt}}) + t_{\mathrm{opt}} \\
&\leq 2t_{\mathrm{opt}}.
\end{aligned}
$$

$\square$

The 2-approximation algorithm for the scheduling problem is adapted from the paper

J. K. Lenstra, D. B. Shmoys, É. Tardos: Approximation algorithms for scheduling unrelated parallel machines, *Mathematical Programming* 46(1990), pages 259–271.

The paper also proves that it is NP-hard to approximate the optimum makespan with a factor less than $\frac{3}{2}$.

Aiming at simplicity, we have presented a somewhat inefficient version of the algorithm. A considerably faster version, with the same approximation guarantee, can be obtained if we do not minimize the function $T \mapsto t^*(T) + T$, but instead we only look for the smallest T with $t^*(T) \leq T$. Such a value of T still guarantees $t^*(T) + T \leq 2t_{\text{opt}}$, but it can be found by binary search over the sorted sequence of the d_{ij}. In this way, it suffices to solve LPR(T) for $O(\log mn)$ values of T, instead of mn values as in our presentation. See the paper quoted above for details.

8.4 Upper Bounds for Codes

Error-correcting codes. Let us consider a DVD player that has a remote control unit with 16 keys. Whenever one of the keys is pressed, the unit needs to communicate this event to the player.

A natural option would be to send a 4-bit sequence: Since there are $2^4 = 16$ different 4-bit sequences (referred to as "words" in this context), a 4-bit sequence is an economical way of communicating one of 16 possibilities.

However, let us suppose that the transmission of bits from the remote control to the player is not quite reliable, and that each of the transmitted bits can be received incorrectly with some small probability, say 0.005. Then we expect that about 2% of the commands are received incorrectly, which can be regarded as a rather serious flaw of the device.

One possibility of improvement is to triple each of the four transmitted bits. That is, instead of a 4-bit word *abcd* the unit sends the 12-bit word *aaabbbcccddd*. Now a transmission error in a *single* bit can be recognized and corrected. For example, if the sequence 111001000111 is received, and if we assume that at most one bit was received erroneously, it is clear that 111000000111 must have been sent. Thus the original 4-bit sequence was 1001. Of course, it might be that actually two or more bits are wrong, and then the original sequence is not reconstructed correctly, but this has much lower probability. Namely, if we assume that the errors in different bits are independent and occur with probability 0.005 (which may or may not be realistic, depending on the technical specifications), then the probability of two or more errors in a 12-bit sequence is approximately 0.16%. This is a significant improvement in reliability. However, the price to pay is transmitting three times as many bits, which presumably exhausts the battery of the remote control much faster.

A significantly better solution to this problem was discovered by Richard Hamming in the 1950s (obviously not in the context of DVD players). In order to distinguish 16 possibilities, we send one of the following 7-bit words: 0000000, 0001011, 0010101, 0011110, 0100110, 0101101, 0110011, 0111000, 1000111, 1001100, 1010010, 1011001, 1100001, 1101010, 1110100, 1111111. It can be checked that every two of these words differ in at least 3 bits. (This fact can be checked by brute force, but it is also a consequence of an elegant general theory, which we do not treat here.) Therefore, if one error occurs in the transmission, the sequence that was sent can be reconstructed uniquely. Hence the capability of correcting any single-bit error is retained, but the number of transmitted bits is reduced to slightly more than half compared to the previous approach.

A similar problem can be investigated for other settings of the parameters as well. In general, we want to communicate one of N possibilities, we do it by transmitting an n-bit word, and we want that any at most r errors can be corrected.

This problem has an enormous theoretical and practical significance. Our example with a DVD player was simple-minded, but error-correcting codes play a role in any technology involving transmission or storage of information, from computer disks and cell phones to deep-space probes. We now introduce some common terminology related to error-correcting codes.

Terminology. The **Hamming distance** of two words $\mathbf{w}, \mathbf{w}' \in \{0,1\}^n$ is the number of bits in which $\mathbf{w}$ differs from $\mathbf{w}'$:

$$d_H(\mathbf{w}, \mathbf{w}') := |\{j \in \{1, \ldots, n\} : w_j \neq w_j'\}|.$$

The Hamming distance can be interpreted as the number of errors "necessary" to transform $\mathbf{w}$ into $\mathbf{w}'$. The **weight** of $\mathbf{w} \in \{0,1\}^n$ is the number of 1's in $\mathbf{w}$:

$$|\mathbf{w}| := |\{j \in \{1, \ldots, n\} : w_j = 1\}|.$$

Finally, for $\mathbf{w}, \mathbf{w}' \in \{0,1\}^n$, we define their sum modulo 2 as the word

$$\mathbf{w} \oplus \mathbf{w}' = ((w_1 + w_1') \bmod 2, \ldots, (w_n + w_n') \bmod 2) \in \{0,1\}^n.$$

These three notions are interrelated by the formula

$$d_H(\mathbf{w}, \mathbf{w}') = |\mathbf{w} \oplus \mathbf{w}'|. \tag{8.6}$$

In the last of the solutions to the DVD-player problem discussed above, the crucial object was the set $\mathcal{C} = \{0000000, 0001011, 0010101, 0011110, 0100110, 0101101, 0110011, 0111000, 1000111, 1001100, 1010010, 1011001, 1100001, 1101010, 1110100, 1111111\}$ of 7-bit words in which every two distinct words had Hamming distance at least 3. In coding theory, any subset $\mathcal{C} \subseteq \{0,1\}^n$ is called a *code*. (This may sound strange, since under a code one usually imagines some kind of method or procedure for coding, but in the theory of error-correcting codes one has to get used to this terminology.) For correcting errors, the crucial parameter is the distance of the code:

8.4.1 Definition. *A code* $C \subseteq \{0,1\}^n$ *has* **distance** d *if* $d_H(\mathbf{w},\mathbf{w}') \geq d$ *for any two distinct words* $\mathbf{w},\mathbf{w}'$ *in* C. *For* $n, d \geq 0$, *let* $A(n,d)$ *denote the maximum cardinality of a code* $C \subseteq \{0,1\}^n$ *with distance* d.

We claim that a code C can correct any at most r errors if and only if it has distance at least $2r+1$. Indeed, on the one hand, if C contained two distinct words $\mathbf{w}', \mathbf{w}''$ that differ in at most $2r$ bits, we consider any word $\mathbf{w}$ resulting from $\mathbf{w}'$ by flipping exactly half of the bits (rounded down) that distinguish $\mathbf{w}'$ from $\mathbf{w}''$. When the word $\mathbf{w}$ is received, there is no way to tell which of the words $\mathbf{w}'$ and $\mathbf{w}''$ was intended. On the other hand, if any two distinct code words differ by at least $2r+1$ bits, then for any word $\mathbf{w} \in \{0,1\}^n$ there is at most one code word from which $\mathbf{w}$ can be obtained through r or fewer errors, and this must be the word that was sent when $\mathbf{w}$ is received.

Given the number n of bits we can afford to transmit and the number r of errors we need to be able to correct, we want a code $C \subseteq \{0,1\}^n$ with distance at least $2r+1$ and with $|C|$ as large as possible, since the number of words in the code corresponds to the amount of information that we can transmit. Thus determining or estimating $A(n,d)$, the maximum possible size of a code $C \subseteq \{0,1\}^n$ with distance d, is one of the main problems of coding theory.

The problem of finding the largest codes for given n and r can *in principle* be solved by complete enumeration: We can go through all possible subsets of $\{0,1\}^n$ and output the largest one that gives a code with distance d. However, this method becomes practically infeasible already for very small n. It turns out that the problem, for arbitrary n and d, is computationally difficult (NP-hard). Starting from very moderate values of n and d, the maximum code sizes are not exactly known, except for few lucky cases. Tightening the known upper and lower bounds on maximum sizes of error-correcting codes is the topic of ongoing research in coding theory.

In this section we present a technique for proving upper bounds based on linear programming. When this technique was introduced by Philippe Delsarte in 1973, it provided upper bounds of unprecedented quality.

Special cases. For all n, we have $A(n,1) = 2^n$, because any code has distance 1. The case $d = 2$ is slightly more interesting. By choosing C as the set of all words of even weight, we see that $A(n,2) \geq 2^{n-1}$. But we actually have $A(n,2) = 2^{n-1}$, since it is easy to show by induction that every code with more than 2^{n-1} words contains two words of Hamming distance 1.

Given the simplicity of the cases $d = 1, 2$, it may come as a surprise that already for $d = 3$, little is known. This is the setup for error-correcting codes with one error allowed. Exact values of $A(n,3)$ have been determined only up to $n \leq 16$; for $n = 17$, for example, the known bounds are

$$5312 \leq A(17,3) \leq 6552.$$

The sphere-packing bound. For any n and d, a simple upper bound on $A(n, d)$ can be obtained by a *volume argument*. Let us motivate this with a real-life analogy. The local grocery is exhibiting a large glass box filled with peas, and the person to make the most accurate estimate of the number of peas in the box wins a prize. Without any counting, you can conclude that the number of peas is bounded above by the volume of the box divided by the volume of a single pea (assuming that all the peas have the same volume).

The same kind of argument can be used for the number $A(n, d)$, where we may assume in our application that $d = 2r + 1$ is odd. Let us fix any code $\mathcal{C}$ of distance d. Now we think of the set $\{0, 1\}^n$ as the glass box, and of the $|\mathcal{C}|$ *Hamming balls*

$$B(\mathbf{w}, r) := \{\mathbf{w}' \in \{0, 1\}^n : d_H(\mathbf{w}, \mathbf{w}') \le r\}, \quad \mathbf{w} \in \mathcal{C},$$

as the peas. Since the code has distance $2r+1$, all these Hamming balls are disjoint and correspond in our analogy to peas. Consequently, their number cannot be larger than the total number of words (the volume of the box) divided by the number of words in a single Hamming ball (the volume of a pea). The number of words at Hamming distance exactly i from $\mathbf{w}$ is $\binom{n}{i}$. This implies

$$|B(\mathbf{w}, r)| = \sum_{i=0}^{r} \binom{n}{i},$$

and the following upper bound on $A(n, 2r + 1)$ is obtained.

8.4.2 Lemma (Sphere-packing bound). *For all n and r,*

$$A(n, 2r + 1) \le \left\lfloor \frac{2^n}{\sum_{i=0}^{r} \binom{n}{i}} \right\rfloor.$$

For example, the sphere-packing bound gives $A(7, 3) \le 16$ (and so the Hamming code in our initial example is optimal), and

$$A(17, 3) \le \lfloor 131072/18 \rfloor = 7281.$$

In the following theorem, which is the main result of this section, an upper bound on $A(n, d)$ is expressed as an optimum of a certain linear program.

8.4.3 Theorem (The Delsarte bound). *For integers n, i, t with $0 \le i$, $t \le n$, let us put*

$$K_t(n, i) = \sum_{j=0}^{\min(i,t)} (-1)^j \binom{i}{j} \binom{n-i}{t-j}.$$

Then for every n and d, $A(n, d)$ is bounded above by the optimum value of the following linear program in variables $x_0, x_1, \ldots, x_n$:

$$
\begin{aligned}
\text{Maximize} \quad & x_0 + x_1 + \cdots + x_n \\
\text{subject to} \quad & x_0 = 1 \\
& x_i = 0, && i = 1, 2, \ldots, d-1 \\
& \sum_{i=0}^{n} K_t(n, i) \cdot x_i \geq 0, && t = 1, 2, \ldots, n \\
& x_0, x_1, \ldots, x_n \geq 0.
\end{aligned}
$$

Example. Using the sphere packing bound, we have previously found $A(17, 3) \leq 7281$. To compute the Delsarte bound, we solve the linear program in the theorem (after eliminating x_0, x_1, x_2, which are actually constants, we have 15 nonnegative variables and 17 constraints). The optimum value is $6553\frac{3}{5}$, which implies $A(17, 3) \leq 6553$. The current best upper bound is 6552, an improvement by only 1!

Toward an explanation. The proof of the Delsarte bound will proceed as follows. With every subset $\mathcal{C} \subseteq \{0, 1\}^n$ we associate nonnegative real quantities $\tilde{x}_0, \tilde{x}_1, \ldots, \tilde{x}_n$ such that $|\mathcal{C}| = \tilde{x}_0 + \cdots + \tilde{x}_n$. Then we will show that whenever $\mathcal{C}$ is a code with distance d, the $\tilde{x}_i$ constitute a feasible solution of the linear program in the theorem. It follows that the maximum of the linear program is at least as large as the size of any existing code $\mathcal{C}$ with distance d (but of course, it may be larger, since a feasible solution does not necessarily corresponds to a code).

Given $\mathcal{C} \subseteq \{0, 1\}^n$, the $\tilde{x}_i$ are defined by

$$
\tilde{x}_i = \frac{1}{|\mathcal{C}|} \cdot \left| \{ (\mathbf{w}, \mathbf{w}') \in \mathcal{C}^2 : d_H(\mathbf{w}, \mathbf{w}') = i \} \right|, \quad i = 0, \ldots, n.
$$

Thus, $\tilde{x}_i$ is the number of ordered pairs of code words with Hamming distance i, divided by the total number of code words. Since any of the $|\mathcal{C}|^2$ ordered pairs contributes to exactly one of the $\tilde{x}_i$, we have

$$
\tilde{x}_0 + \tilde{x}_1 + \cdots + \tilde{x}_n = |\mathcal{C}|.
$$

Some of the constraints in the linear program in Theorem 8.4.3 are now easy to understand. We clearly have $\tilde{x}_0 = 1$, since every $\mathbf{w} \in \mathcal{C}$ has distance 0 only to itself. The equations $\tilde{x}_1 = 0$ through $\tilde{x}_{d-1} = 0$ hold by the assumption that $\mathcal{C}$ has distance d; that is, there are no pairs of code words with Hamming distance between 1 and $d-1$. Interestingly, this is the *only* place in the proof of the Delsarte bound where the assumption of $\mathcal{C}$ being a code with distance d is used.

The remaining set of constraints is considerably harder to derive, and it lacks a really intuitive explanation. Thus, to prove Theorem 8.4.3, we have to establish the following.

8.4.4 Proposition. *Let $\mathcal{C} \subseteq \{0, 1\}^n$ be an arbitrary set, let $\tilde{x}_i = \tilde{x}_i(\mathcal{C})$ be defined as above, and let $t \in \{1, 2, \ldots, n\}$. Then we have the inequality*

$$\sum_{i=0}^{n} K_t(n, i) \cdot \tilde{x}_i \geq 0.$$

In the next lemma, $I \subseteq \{1, 2, \ldots, n\}$ is a set of indices, and we write $d_H^I(\mathbf{w}, \mathbf{w}')$ for the number of indices $i \in I$ with $w_i \neq w_i'$ (thus, the components outside I are ignored).

8.4.5 Lemma. *Let $I \subseteq \{1, 2, \ldots, n\}$ be a set of indices, and let $\mathcal{C} \subseteq \{0, 1\}^n$. Then the number of pairs $(\mathbf{w}, \mathbf{w}') \in \mathcal{C}^2$ with $d_H^I(\mathbf{w}, \mathbf{w}')$ even is at least as large as the number of pairs $(\mathbf{w}, \mathbf{w}') \in \mathcal{C}^2$ with $d_H^I(\mathbf{w}, \mathbf{w}')$ odd. (In probabilistic terms, if we choose $\mathbf{w}, \mathbf{w}' \in \mathcal{C}$ independently at random, then the probability that they differ in an even number of positions from I is at least as large as the probability that they differ in an odd number of positions from I.)*

Proof. Let us write $|\mathbf{w}|_I = |\{i \in I : w_i = 1\}|$, and let us set

$$\mathcal{E} = \{\mathbf{w} \in \mathcal{C} : |\mathbf{w}|_I \text{ is even}\}, \quad \mathcal{O} = \{\mathbf{w} \in \mathcal{C} : |\mathbf{w}|_I \text{ is odd}\}.$$

From the equation $d_H^I(\mathbf{w}, \mathbf{w}') = |\mathbf{w} \oplus \mathbf{w}'|_I$, we see that if $d_H^I(\mathbf{w}, \mathbf{w}')$ is even, then $|\mathbf{w}|_I$ and $|\mathbf{w}'|_I$ have the same parity, and so $\mathbf{w}$ and $\mathbf{w}'$ are both in $\mathcal{E}$ or both in $\mathcal{O}$. On the other hand, for $d_H^I(\mathbf{w}, \mathbf{w}')$ odd, one of $\mathbf{w}, \mathbf{w}'$ lies in $\mathcal{E}$ and the other one in $\mathcal{O}$. So the assertion of the lemma is equivalent to $|\mathcal{E}|^2 + |\mathcal{O}|^2 \geq 2 \cdot |\mathcal{E}| \cdot |\mathcal{O}|$, which follows by expanding $(|\mathcal{E}| - |\mathcal{O}|)^2 \geq 0$. □

8.4.6 Corollary. *For every $\mathcal{C} \subseteq \{0, 1\}^n$ and every $\mathbf{v} \in \{0, 1\}^n$ we have*

$$\sum_{(\mathbf{w}, \mathbf{w}') \in \mathcal{C}^2} (-1)^{(\mathbf{w} \oplus \mathbf{w}')^T \mathbf{v}} \geq 0.$$

Proof. This is just another way of writing the statement of Lemma 8.4.5. Indeed, if we set $I = \{i : v_i = 1\}$, then $(\mathbf{w} \oplus \mathbf{w}')^T \mathbf{v} = d_H^I(\mathbf{w}, \mathbf{w}')$, and hence the sum in the corollary is exactly the number of pairs $(\mathbf{w}, \mathbf{w}')$ with $d_H^I(\mathbf{w}, \mathbf{w}')$ even minus the number of pairs with $d_H^I(\mathbf{w}, \mathbf{w}')$ odd.

The corollary also has a quick algebraic proof, which some readers may prefer. It suffices to note that $(\mathbf{w} \oplus \mathbf{w}')^T \mathbf{v}$ has the same parity as $(\mathbf{w} + \mathbf{w}')^T \mathbf{v}$ (addition modulo 2 was replaced by ordinary addition of integers), and so

$$\sum_{(\mathbf{w}, \mathbf{w}') \in \mathcal{C}^2} (-1)^{(\mathbf{w} \oplus \mathbf{w}')^T \mathbf{v}} = \sum_{(\mathbf{w}, \mathbf{w}') \in \mathcal{C}^2} (-1)^{(\mathbf{w} + \mathbf{w}')^T \mathbf{v}}$$

$$= \sum_{(\mathbf{w}, \mathbf{w}') \in \mathcal{C}^2} (-1)^{\mathbf{w}^T \mathbf{v}} \cdot (-1)^{\mathbf{w}'^T \mathbf{v}}$$

$$= \left(\sum_{\mathbf{w} \in \mathcal{C}} (-1)^{\mathbf{w}^T \mathbf{v}} \right)^2 \geq 0.$$

□

Proof of Proposition 8.4.4. To prove the tth inequality in the proposition, i.e., $\sum_{i=0}^{n} K_t(n,i) \cdot \tilde{x}_i \geq 0$, we sum the inequality in Corollary 8.4.6 over all $\mathbf{v} \in \{0,1\}^n$ of weight t. Interchanging the summation order, we obtain

$$0 \leq \sum_{(\mathbf{w},\mathbf{w}') \in \mathcal{C}^2} \sum_{\mathbf{v} \in \{0,1\}^n:\, |\mathbf{v}|=t} (-1)^{(\mathbf{w} \oplus \mathbf{w}')^T \mathbf{v}}.$$

To understand this last expression, let us fix $\mathbf{u} = \mathbf{w} \oplus \mathbf{w}'$ and write $S(\mathbf{u}) = \sum_{\mathbf{v} \in \{0,1\}^n:\, |\mathbf{v}|=t} (-1)^{\mathbf{u}^T \mathbf{v}}$. In this sum, the $\mathbf{v}$ with $\mathbf{u}^T \mathbf{v} = j$ are counted with sign $(-1)^j$. How many $\mathbf{v}$ of weight t and with $\mathbf{u}^T \mathbf{v} = j$ are there? Let $i = |\mathbf{u}| = d_H(\mathbf{w}, \mathbf{w}')$ be the number of 1's in $\mathbf{u}$. In order to form such a $\mathbf{v}$, we need to put j ones in positions where $\mathbf{u}$ has 1's and $t - j$ ones in positions where $\mathbf{u}$ has 0's. Hence the number of these $\mathbf{v}$ is $\binom{i}{j}\binom{n-i}{t-j}$, and

$$S(\mathbf{u}) = \sum_{j=0}^{\min(i,t)} (-1)^j \binom{i}{j}\binom{n-i}{t-j},$$

which we recognize as $K_t(n,i)$. So we have

$$0 \leq \sum_{(\mathbf{w},\mathbf{w}') \in \mathcal{C}^2} K_t(n, d_H(\mathbf{w},\mathbf{w}')),$$

and it remains to note that the number of times $K_t(n,i)$ appears in this sum is $|\mathcal{C}| \cdot \tilde{x}_i$. This finishes the proof of Proposition 8.4.4 and thus also of Theorem 8.4.3. $\qquad\square$

A small strengthening of the Delsarte bound. We have seen that Theorem 8.4.3 yields $A(17,3) \leq 6553$. We show how the inequalities in the theorem can be slightly strengthened using a parity argument, which leads to the best known upper bound $A(17,3) \leq 6552$. Similar tricks can improve the Delsarte bound in some other cases as well, but the improvements are usually minor.

For contradiction let us suppose that $n = 17$ and there is a code $\mathcal{C} \subseteq \{0,1\}^n$ of distance 3 with $|\mathcal{C}| = 6553$. The size of $\mathcal{C}$ is odd, and we note that for every code of odd size and every t, the last inequality in the proof of Corollary 8.4.6 can be strengthened to

$$\left(\sum_{\mathbf{w} \in \mathcal{C}} (-1)^{\mathbf{w}^T \mathbf{v}} \right)^2 \geq 1,$$

since an odd number of values from $\{-1,1\}$ cannot sum to zero. If we propagate this improvement through the proof of Proposition 8.4.4, we arrive at the following inequality for the $\tilde{x}_i$:

$$\sum_{i=0}^{n} K_t(n,i) \cdot \tilde{x}_i \geq \frac{\binom{n}{t}}{|\mathcal{C}|} .$$

Since in our particular case we suppose $|\mathcal{C}| = 6553$, we can replace the constraints $\sum_{i=0}^{n} K_t(n,i) \cdot x_i \geq 0$, $t = 1, 2, \ldots, n$, in the linear program in Theorem 8.4.3 by $\sum_{i=0}^{n} K_t(n,i) \cdot x_i \geq \binom{n}{t}/6553$, and the $\tilde{x}_i$ defined by our $\mathcal{C}$ remain a feasible solution. However, the optimum of this modified linear program is only $6552\frac{3}{5}$, which contradicts the assumption $|\mathcal{C}| = 6553$. This proves $A(17,3) \leq 6552$.

The paper

M. R. Best, A. E. Brouwer, F. J. MacWilliams, A. M. Odlyzko, and N. J. A. Sloane: Bounds for binary codes of length less than 25, *IEEE Trans. Inform. Theory* 24 (1978), pages 81–93.

describes this particular strengthening of the Delsarte bound and some similar approaches. A continually updated table of the best known bounds for $A(n, d)$ for small n and d is maintained by Andries Brouwer at

http://www.win.tue.nl/~aeb/codes/binary-1.html.

The Delsarte bound explained. The result in Theorem 8.4.3 goes back to the thesis of Philippe Delsarte:

P. Delsarte: An algebraic approach to the association schemes of coding theory, *Philips Res. Repts. Suppl.* 10 (1973).

Here we sketch Delsarte's original proof. At a comparable level of detail, his proof is more involved than the ad hoc proof above (from the paper by Best et al.). On the other hand, Delsarte's proof is more systematic, and even more importantly, it can be extended to prove a stronger result, which we mention below.

For $i \in \{0, \ldots, n\}$, let M_i be the $2^n \times 2^n$ matrix defined by[1]

$$(M_i)_{\mathbf{v},\mathbf{w}} = \begin{cases} 1 & \text{if } d_H(\mathbf{v}, \mathbf{w}) = i \\ 0 & \text{otherwise.} \end{cases}$$

The set of matrices of the form

$$\sum_{i=0}^{n} y_i M_i, \quad y_0, \ldots, y_n \in \mathbb{R},$$

is known to be closed under addition and scalar multiplication (this is clear), and under matrix multiplication (this has to be shown). A set

---

[1] We assume that rows and columns are indexed by the words from $\{0,1\}^n$.

of matrices closed under these operations is called a **matrix algebra**. In our case, one speaks of the **Bose–Mesner algebra of the Hamming association scheme**. The matrix multiplication turns out to be commutative on this algebra, and this is known to imply a strong condition: The M_i have a *common* diagonalization, meaning that there is an orthogonal matrix U with $U^T M_i U$ diagonal for all i.

Once we know this, it is a matter of patience to find such a matrix U. For example, the matrix U defined by

$$U_{\mathbf{v},\mathbf{w}} = \frac{1}{2^{n/2}}(-1)^{\mathbf{v}^T\mathbf{w}}$$

will do. First we have to check (this is easy) that this matrix is indeed orthogonal, meaning that $U^T U = I_n$.

For the entries of $U^T M_i U$, we can derive the formula

$$(U^T M_i U)_{\mathbf{v},\mathbf{w}} = \frac{1}{2^n} \sum_{\substack{(\mathbf{u},\mathbf{u}')\in(\{0,1\}^n)^2 \\ d_H(\mathbf{u},\mathbf{u}')=i}} (-1)^{\mathbf{u}^T\mathbf{v}+\mathbf{u}'^T\mathbf{w}}.$$

We claim that this sum evaluates to 0 whenever $\mathbf{v} \neq \mathbf{w}$; this will imply that $U^T M_i U$ is indeed a diagonal matrix. To prove the claim, we let j be any index for which $v_j \neq w_j$. In the sum, we can then pair up the terms for $(\mathbf{u},\mathbf{u}')$ and $(\mathbf{u}\oplus\mathbf{e}_j, \mathbf{u}'\oplus\mathbf{e}_j)$, with $\mathbf{e}_j$ being the word with a 1 exactly at position j. This pairing covers all terms of the sum, and paired-up terms are easily seen to cancel each other. (If you didn't believe $U^T U = I_n$, this can be shown with an even simpler pairing argument along these lines.)

On the diagonal, we get

$$(U^T M_i U)_{\mathbf{w},\mathbf{w}} = \frac{1}{2^n} \sum_{\substack{(\mathbf{u},\mathbf{u}')\in(\{0,1\}^n)^2 \\ d_H(\mathbf{u},\mathbf{u}')=i}} (-1)^{(\mathbf{u}\oplus\mathbf{u}')^T\mathbf{w}}$$

$$= \sum_{\substack{\mathbf{v}\in\{0,1\}^n \\ |\mathbf{v}|=i}} (-1)^{\mathbf{v}^T\mathbf{w}} = K_i(n,|\mathbf{w}|)$$

(for the last equality see the proof of Proposition 8.4.4) since any $\mathbf{v}$ of weight i can be written in the form $\mathbf{v} = \mathbf{u} \oplus \mathbf{u}'$ in 2^n different ways, one for each $\mathbf{u} \in \{0,1\}^n$.

Next, let us fix a code $\mathcal{C}$ and look at a specific matrix in the Bose–Mesner algebra. For this, we define the values

$$\tilde{y}_i = \frac{|\{(\mathbf{w},\mathbf{w}') \in \mathcal{C}^2 : d_H(\mathbf{w},\mathbf{w}') = i\}|}{2^n\binom{n}{i}}, \quad i = 0,\ldots,n.$$

We note that $\tilde{y}_i$ is the probability that a randomly chosen pair of words with Hamming distance i is a pair of code words. Moreover, $\tilde{y}_i$ is related to our earlier quantity $\tilde{x}_i$ via

$$\tilde{y}_i = \frac{|\mathcal{C}|}{2^n \binom{n}{i}} \tilde{x}_i. \tag{8.7}$$

Here comes Delsarte's main insight.

8.4.7 Lemma. *The matrix* $\tilde{M} = \sum_{i=0}^n \tilde{y}_i M_i$ *is positive semidefinite.*

Proof. We first observe that

$$\tilde{M}_{\mathbf{v},\mathbf{w}} = \tilde{y}_i, \tag{8.8}$$

where $i = d_H(\mathbf{v},\mathbf{w})$. We will express $\tilde{M}$ as a positive linear combination of matrices that are obviously positive semidefinite. We start with the matrix $X^{\mathcal{C}}$ defined by

$$X^{\mathcal{C}}_{\mathbf{v},\mathbf{w}} = \begin{cases} 1 & \text{if } (\mathbf{v},\mathbf{w}) \in \mathcal{C}^2 \\ 0 & \text{otherwise.} \end{cases}$$

This matrix is positive semidefinite, since it can be written in the form

$$X^{\mathcal{C}} = \mathbf{x}^{\mathcal{C}}(\mathbf{x}^{\mathcal{C}})^T,$$

where $\mathbf{x}^{\mathcal{C}}$ is the characteristic vector of $\mathcal{C}$:

$$\mathbf{x}^{\mathcal{C}}_{\mathbf{w}} := \begin{cases} 1 & \text{if } \mathbf{w} \in \mathcal{C} \\ 0 & \text{otherwise.} \end{cases}$$

Let Π be the automorphism group of $\{0,1\}^n$, consisting of the $n!2^n$ bijections that permute indices and swap 0's with 1's at selected positions. With τ chosen uniformly at random from Π, we obtain[2]

$$\text{Prob}\Big[(\mathbf{v},\mathbf{w}) \in \tau(\mathcal{C})^2\Big] = \frac{1}{|\Pi|} \sum_{\pi \in \Pi} \Big[(\mathbf{v},\mathbf{w}) \in \pi(\mathcal{C})^2\Big]$$

$$= \frac{1}{|\Pi|} \sum_{\pi \in \Pi} X^{\pi(\mathcal{C})}_{\mathbf{v},\mathbf{w}}.$$

On the other hand,

$$\text{Prob}\Big[(\mathbf{v},\mathbf{w}) \in \tau(\mathcal{C})^2\Big] = \text{Prob}\Big[(\tau^{-1}(\mathbf{v}), \tau^{-1}(\mathbf{w})) \in \mathcal{C}^2\Big] = \tilde{y}_i,$$

since τ^{-1} is easily shown to map $(\mathbf{v},\mathbf{w})$ to a *random* pair of words with Hamming distance i.

---

[2] The *indicator variable* $[P]$ of a statement P has value 1 if P holds and 0 otherwise.

Using (8.8), this shows that

$$\tilde{M} = \frac{1}{|\Pi|} \sum_{\pi \in \Pi} X^{\pi(\mathcal{C})}$$

is a positive linear combination of positive semidefinite matrices and is therefore positive semidefinite itself. The lemma is proved.   □

After diagonalization of $\tilde{M}$ by the matrix U, the statement of Lemma 8.4.7 can equivalently be written as

$$\sum_{i=0}^{n} \tilde{y}_i (U^T M_i U)_{\mathbf{w},\mathbf{w}} = \sum_{i=0}^{n} \tilde{y}_i K_i(n, |\mathbf{w}|) \geq 0, \quad \mathbf{w} \in \{0,1\}^n.$$

This is true since diagonalization preserves the property of being positive semidefinite, which for diagonal matrices is equivalent to nonnegativity of all diagonal entries.

Taking into account the relation (8.7) between the $\tilde{y}_i$ and our original $\tilde{x}_i$, this implies the following inequalities for any code $\mathcal{C}$.

$$\sum_{i=0}^{n} \tilde{x}_i \frac{K_i(n, t)}{\binom{n}{i}} \geq 0, \quad t = 1, \ldots, n. \tag{8.9}$$

Observing that

$$\frac{\binom{t}{j} \binom{n-t}{i-j}}{\binom{n}{i}} = \frac{\binom{i}{j} \binom{n-i}{t-j}}{\binom{n}{t}},$$

we get (cf. the definition of K_t in Theorem 8.4.3)

$$\frac{K_i(n, t)}{\binom{n}{i}} = \frac{K_t(n, i)}{\binom{n}{t}}.$$

Under this equation, the inequalities in (8.9) are equivalent to those in Proposition 8.4.4 and we recover the Delsarte bound.

Beyond the Delsarte bound. Alexander Schrijver generalized Delsarte's approach and improved the upper bounds on $A(n, d)$ significantly in many cases:

A. Schrijver: New code upper bounds from the Terwilliger algebra and semidefinite programming, *IEEE Trans. Inform. Theory* 51 (2005), pages 2859–2866.

His work uses semidefinite programming instead of linear programming.

8.5 Sparse Solutions of Linear Systems

A coding problem. We begin with discussing error-correcting codes again, but this time we want to send a sequence $w \in \mathbb{R}^k$ of k real numbers. Or rather not we, but a deep-space probe which needs to transmit its priceless measurements represented by w back to Earth. We want to make sure that all components of w can be recovered correctly even if some fraction, say 8%, of the transmitted numbers are corrupted, due to random errors or even maliciously (imagine that the secret *Brotherhood for Promoting the Only Truth* can somehow tamper with the signal slightly in order to document the presence of supernatural phenomena in outer space). We admit *gross errors*; that is, if the number 3.1415 is sent and it gets corrupted, it can be received as 2152.66, or 3.1425, or -10^{11}, or any other real number.

Here is a way of encoding w: We choose a suitable number $n > k$ and a suitable $n \times k$ *encoding matrix* Q of rank k, and we send the vector $z = Qw \in \mathbb{R}^n$. Because of the errors, the received vector is not z but $\tilde{z} = z + x$, where $x \in \mathbb{R}^n$ is a vector with at most $r = \lfloor 0.08n \rfloor$ nonzero components. We ask, under what conditions on Q can z be recovered from $\tilde{z}$?

Somewhat counterintuitively, we will concentrate on the task of finding the "error vector" x. Indeed, once we know x, we can compute w by solving the system of linear equations $Qw = z = \tilde{z} - x$. The solution, if one exists, is unique, since we assume that Q has rank k and hence the mapping $w \mapsto Qw$ is injective.

Sparse solutions of underdetermined linear systems. In order to compute x, we first reformulate the recovery problem. Let $m = n - k$ and let A be an $m \times n$ matrix such that $AQ = 0$. That is, considering the k-dimensional linear subspace of $\mathbb{R}^n$ generated by the columns of Q, the rows of A form a basis of its orthogonal complement. The following picture illustrates the dimensions of the matrices:

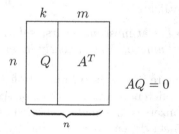

In the recovery problem we have $\tilde{z} = Qw + x$. Multiplying both sides by A from the left, we obtain $A\tilde{z} = AQw + Ax = Ax$. Setting $b = A\tilde{z}$, we thus get that the unknown x has to satisfy the system of linear equations $Ax = b$. We have $m = n - k$ equations and $n > m$ unknowns; the system is *underdetermined* and it has infinitely many solutions. In general, not all of these solutions can appear as an error vector in the decoding problem (we note that the multiplication by A above is not necessarily an equivalent transformation and so it may give rise to spurious solutions). However, we seek a solution $\mathbf{x}$ with the extra property $|\mathrm{supp}(\mathbf{x})| \leq r$, where we introduce the notation

$$\mathrm{supp}(\mathbf{x}) = \{i \in \{1, 2, \ldots, n\} : x_i \neq 0\}.$$

As we will see, under suitable conditions relating n, m, r and A, such a *sparse* solution of $A\mathbf{x} = \mathbf{b}$ turns out to be unique (and thus it has to be the desired error vector), and it can be computed efficiently by linear programming!

Let us summarize the resulting problem once again:

Sparse solution of underdetermined system of linear equations

Given an $m{\times}n$ matrix A with $m < n$, a vector $\mathbf{b} \in \mathbb{R}^m$, and an integer r, find an $\mathbf{x} \in \mathbb{R}^n$ such that

$$A\mathbf{x} = \mathbf{b} \quad \text{and} \quad |\mathrm{supp}(\mathbf{x})| \leq r \qquad (8.10)$$

if one exists.

The coding problem above is only one among several important practical problems leading to the computation of sparse solutions of underdetermined systems. We will mention other applications at the end of this section. From now on, we call any $\mathbf{x}$ satisfying (8.10) a **sparse solution** to $A\mathbf{x} = \mathbf{b}$. (Warning: This shouldn't be confused with solutions of *sparse systems* of equations, which is an even more popular topic in numerical mathematics and scientific computing.)

A linear algebra view. There is a simple necessary and sufficient condition guaranteeing that there is at most one sparse solution of $A\mathbf{x} = \mathbf{b}$.

8.5.1 Observation. *With n, m, r fixed, the following two conditions on an $m{\times}n$ matrix A are equivalent:*

(i) *The system $A\mathbf{x} = \mathbf{b}$ has at most one sparse solution $\mathbf{x}$ for every $\mathbf{b}$.*
(ii) *Every $2r$ or fewer columns of A are linearly independent.*

Proof. To prove the (more interesting) implication (ii)$\Rightarrow$(i), let us assume that $\mathbf{x}'$ and $\mathbf{x}''$ are two different sparse solutions of $A\mathbf{x} = \mathbf{b}$. Then $\mathbf{y} = \mathbf{x}' - \mathbf{x}'' \neq \mathbf{0}$ has at most $2r$ nonzero components and satisfies $A\mathbf{y} = A\mathbf{x}' - A\mathbf{x}'' = \mathbf{0}$, and hence it defines a linear dependence of at most $2r$ columns of A.

To prove (i)$\Rightarrow$(ii), we essentially reverse the above argument. Supposing that there exists nonzero $\mathbf{y} \in \mathbb{R}^n$ with $A\mathbf{y} = \mathbf{0}$ and $|\mathrm{supp}(\mathbf{y})| \leq 2r$, we write $\mathbf{y} = \mathbf{x}' - \mathbf{x}''$, where both $\mathbf{x}'$ and $\mathbf{x}''$ have at most r nonzero components. For example, $\mathbf{x}'$ may agree with $\mathbf{y}$ in the first $\lfloor |\mathrm{supp}(\mathbf{y})|/2 \rfloor$ nonzero components and have 0's elsewhere, and $\mathbf{x}'' = \mathbf{x}' - \mathbf{y}$ has the remaining at most r nonzero components of $\mathbf{y}$ with opposite sign. We set $\mathbf{b} = A\mathbf{x}'$, so that $\mathbf{x}'$ is a sparse solution of $Ax = b$, and we note that x'' is another sparse solution since $Ax'' = Ax' - Ay = Ax' = b$.   □

Let us note that (ii) implies that, in particular, $m \geq 2r$. On the other hand, if we choose a "random" $2r \times n$ matrix A, we almost surely have every $2r$ columns linearly independent.[3] So in the coding problem, if we set n so that $n = k + 2r$, choose A randomly, and let the columns of Q form a basis of the orthogonal complement of the row space of A, we seem to be done—a random A has almost surely every $2r$ columns linearly independent, and in such case, assuming that no more than r errors occurred, the sparse error vector x is always determined uniquely, and so is the original message w.

Efficiency? But a major question remains—how can we *find* the unknown sparse solution x? Unfortunately, it turns out that the problem of computing a sparse solution of $Ax = b$ is difficult (NP-hard) in general, even for A satisfying the conditions of Observation 8.5.1.

Since the problem of finding a sparse solution of $Ax = b$ is important and computationally difficult, several heuristic methods have been proposed for solving it at least approximately and at least in some cases. One of them, described next, turned out to be considerably more powerful than the others.

Basis pursuit. A sparse solution x is "small" in the sense of having few nonzero components. The idea is to look for x that is "small" in another sense that is easier to deal with, namely, with small $|x_1| + |x_2| + \cdots + |x_n|$. The last quantity is commonly denoted by $\|x\|_1$ and called the ℓ_1-**norm** of x (while $\|x\| = \|x\|_2 = \sqrt{x_1^2 + \cdots + x_n^2}$ is the usual Euclidean norm, which can also be called the ℓ_2-norm).[4] We thus arrive at the following optimization problem (usually called *basis pursuit* in the literature):

$$\text{Minimize } \|x\|_1 \text{ subject to } x \in \mathbb{R}^n \text{ and } Ax = b. \tag{BP}$$

---

[3] In this book we don't want to assume or introduce the knowledge required to state and prove this claim rigorously. Instead, we offer the following semiformal argument relying on a famous but nontrivial theorem. The condition of linear independence of every $2r$ columns can be reformulated as $\det(A_I) \neq 0$ for every $2r$-element $I \subset \{1, 2, \ldots, n\}$. Now for I fixed, $\det(A_I)$ is a polynomial of degree $2r$ in the $2rn$ entries of A (it really depends only on $4r^2$ entries but never mind), and the set of the matrices A with $\det(A_I) = 0$ is the zero set of this polynomial in $\mathbb{R}^{2rn}$. The zero set of any nonzero polynomial is very "thin"; by Sard's theorem, it has Lebesgue measure 0. Hence the matrices A with $\det(A_I) = 0$ for at least one I correspond to points in $\mathbb{R}^{2nr}$ lying on the union of $\binom{n}{2r}$ zero sets, each of measure 0, and altogether they have measure 0. Therefore, such matrices appear with zero probability in any "reasonable" continuous distribution on matrices, for example, if the entries of A are chosen independently and uniformly from the interval $[-1, 1]$.

[4] The letter ℓ here can be traced back to the surname of Henri Lebesgue, the founder of modern integration theory. A certain space of integrable functions on $[0, 1]$ is denoted by $L_1(0, 1)$ in his honor, and ℓ_1 is a "pocket version" of this space consisting of countable sequences instead of functions.

By a trick we have learned in Section 2.4, this problem can be reformulated as a linear program:

$$\begin{array}{ll} \text{Minimize} & u_1 + u_2 + \cdots + u_n \\ \text{subject to} & A\mathbf{x} = \mathbf{b} \\ & -\mathbf{u} \le \mathbf{x} \le \mathbf{u} \\ & \mathbf{x}, \mathbf{u} \in \mathbb{R}^n, \ \mathbf{u} \ge \mathbf{0}. \end{array} \qquad \text{(BP')}$$

To check the equivalence of (BP) and (BP'), we just note that in an optimal solution of (BP') we have $u_i = |x_i|$ for every i.

The basis pursuit approach to finding a sparse solution of $A\mathbf{x} = \mathbf{b}$ thus consists in computing an optimal solution $\mathbf{x}^*$ of (BP) by linear programming, and hoping that, with some luck, this $\mathbf{x}^*$ might also be the sparse solution or at least close to it.

At first sight it is not clear why basis pursuit should have any chance of finding a sparse solution. After all, the desired sparse solution might have a few huge components, while $\mathbf{x}^*$, a minimizer of the ℓ_1-norm, might have all components nonzero but tiny.

Surprisingly, experiments have revealed that basis pursuit actually performs excellently, and it usually finds the sparse solution exactly even in conditions that don't look very favorable. Later these findings were confirmed by rather precise theoretical results. Here we state the following particular case of such results:

8.5.2 Theorem (Guaranteed success of basis pursuit). *Let*

$$m = \lfloor 0.75n \rfloor,$$

and let A be a random $m \times n$ matrix, where each entry is drawn from the standard normal distribution $N(0,1)$ and the entries are mutually independent.[5] Then with probability at least $1 - e^{-cm}$, where $c > 0$ is a positive constant, the matrix A has the following property:

> *If $\mathbf{b} \in \mathbb{R}^m$ is such that the system $A\mathbf{x} = \mathbf{b}$ has a solution $\tilde{\mathbf{x}}$ with at most $r = \lfloor 0.08n \rfloor$ nonzero components, then $\tilde{\mathbf{x}}$ is a unique optimal solution of (BP).*

For brevity, we call a matrix A with the property as in the theorem **BP-exact** (more precisely, we should say "BP-exact for r," where r specifies the maximum number of nonzero components). For a BP-exact matrix A we can

---

[5] We recall that the distribution $N(0,1)$ has density given by the Gaussian "bell curve" $\frac{1}{\sqrt{2\pi}} e^{-x^2/2}$. How can we generate a random number with this distribution? This is implemented in many software packages, and methods for doing it can be found, for instance, in

> D. Knuth: *The Art of Computer Programming*, Vol. 2: Seminumerical Algorithms, Addison-Wesley, Reading, Massachusetts, 1973.

thus find a sparse solution of $A\mathbf{x} = \mathbf{b}$ exactly and efficiently, by solving the linear program (BP').

Returning to the coding problem from the beginning of the section, we immediately obtain the following statement:

8.5.3 Corollary. *Let k be a sufficiently large integer, let us set $n = 4k$, $m = 3k$, let a random $m{\times}n$ matrix A be generated as in Theorem 8.5.2, and let Q be an $n{\times}k$ matrix of rank k with $AQ = 0$ (in other words, the column space of Q is the orthogonal complement of the row space of A). Then the following holds with probability overwhelmingly close to 1: If Q is used as a coding matrix to transmit a vector $\mathbf{w} \in \mathbb{R}^k$, by sending the vector $\mathbf{z} = Q\mathbf{w} \in \mathbb{R}^n$, then even if any at most 8% of the entries of $\mathbf{z}$ are corrupted, we can still reconstruct $\mathbf{w}$ exactly and efficiently, by solving the appropriate instance of (BP').*

Drawing the elements of A from the standard normal distribution is not the only known way of generating a BP-exact matrix. Results similar to Theorem 8.5.2, perhaps with worse constants, can be proved by known techniques for random matrices with other distributions. The perhaps simplest such distribution is obtained by choosing each entry to be $+1$ or -1, each with probability $\frac{1}{2}$ (and again with all entries mutually independent).

A somewhat unpleasant feature of Theorem 8.5.2 and of similar results is that they provide a BP-exact matrix only with high probability. No efficient method for *verifying* that a given matrix is BP-exact is known at present, and so we cannot be absolutely sure. In practice this is not really a problem, since the probability of failure (i.e., of generating a matrix that is not BP-exact) can be made very small by choosing the parameters appropriately, much smaller than the probability of sudden death of all people in the team that wants to compute the sparse solution, for instance. Still, it would be nice to have explicit constructions of BP-exact matrices with good parameters.

Random errors. In our coding problem, we allow for completely arbitrary (even malicious) errors; all we need is that there aren't too many errors. However, in practice one may often assume that the errors occur at random positions, and we want to be able to decode correctly only with high probability, that is, for most (say 99.999%) of the $\binom{n}{r}$ possible locations of the r errors. It turns out that considerably stronger numerical bounds can be obtained in this setting: For example, Theorem 8.5.2 tells us that for $m = \lfloor 0.75n \rfloor$ and $k = n - m$, we are guaranteed to fix *any* $0.08n$ errors with high probability, but it turns out that we can also fix *most* of the possible r-tuples of errors for r as large as $0.36n$! For a precise statement see the paper of Donoho quoted near the end of the section.

Geometric meaning of BP-exactness. Known proofs of Theorem 8.5.2 or similar results use a good deal of geometric probability and high-dimensional geometry, knowledge which we want neither to assume nor to introduce in this book. We thus have to omit a proof. Instead, we present an appealing geometric characterization of BP-exact matrices, which is a starting point of existing proofs.

For its statement we need to recall the *crosspolytope*, a convex polytope already mentioned in Section 4.3. We will denote the n-dimensional crosspolytope by B_1^n, which should suggest that it is the unit ball of the ℓ_1-norm:

$$B_1^n = \{\mathbf{x} \in \mathbb{R}^n : \|\mathbf{x}\|_1 \leq 1\}.$$

8.5.4 Lemma (Reformulation of BP-exactness). *Let A be an $m \times n$ matrix, $m < n$, let $r \leq m$, and let $L = \{\mathbf{x} \in \mathbb{R}^n : A\mathbf{x} = \mathbf{0}\}$ be the kernel (null space) of A. Then A is BP-exact for r if and only if the following holds: For every $\mathbf{z} \in \mathbb{R}^n$ with $\|\mathbf{z}\|_1 = 1$ (i. e., $\mathbf{z}$ is a boundary point of the crosspolytope) and $|\mathrm{supp}(\mathbf{z})| \leq r$, the affine subspace $L + \mathbf{z}$ intersects the crosspolytope only at $\mathbf{z}$; that is, $(L + \mathbf{z}) \cap B_1^n = \{\mathbf{z}\}$.*

Let us discuss an example with $n = 3$, $m = 2$, and $r = 1$, about the only sensible setting for which one can draw a picture. If the considered 2×3 matrix A has full rank, which we may assume, then the kernel L is a one-dimensional linear subspace of $\mathbb{R}^3$, that is, a line passing through the origin. The points $\mathbf{z}$ coming into consideration have at most $r = 1$ nonzero coordinate, and they lie on the boundary of the regular octahedron B_1^3, and hence they are precisely the 6 vertices of B_1^3:

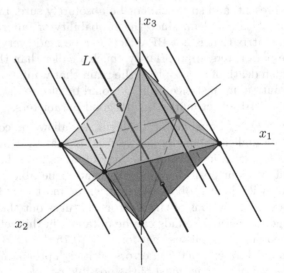

The line L through the origin is drawn thick, and the condition in the lemma says that each of the 6 translates of L to the vertices should only touch the crosspolytope. Another way of visualizing this is to look at the projection of B_1^3 to the plane orthogonal to L. Each of the translates of L is projected to a point, and the crosspolytope is projected to a convex polygon. The condition then means that all the 6 vertices should appear on the boundary in the projection, as in the left picture below,

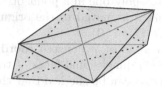 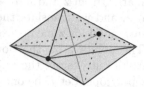

while the right picture corresponds to a bad L (the condition is violated at the two vertices marked by dots that lie inside the projection). In general, of course, L is not a line but a k-dimensional linear subspace of $\mathbb{R}^n$, and the considered points $\mathbf{z}$ are not only vertices of B_1^n, but they can lie in all $(r-1)$-dimensional faces of B_1^n. Indeed, we note that the points $\mathbf{z}$ on the surface of B_1^n with at most r nonzero components are exactly the points of the union of all $(r-1)$-dimensional faces, omitting the easy proof of this fact (but look at least at the case $n = 3$, $r = 2$).

Proof of Lemma 8.5.4. First we assume that A is BP-exact, we consider a point $\mathbf{z}$ with $\|\mathbf{z}\|_1 = 1$ and $|\mathrm{supp}(\mathbf{z})| \leq r$, and we set $\mathbf{b} = A\mathbf{z}$. Then the system $A\mathbf{x} = \mathbf{b}$ has a sparse solution, namely $\mathbf{z}$, and hence $\mathbf{z}$ has to be the unique point among all solutions of $A\mathbf{x} = \mathbf{b}$ that minimize the ℓ_1-norm. Noting that the set of all solutions of $A\mathbf{x} = \mathbf{b}$ is exactly the affine subspace $L + \mathbf{z}$, we get that $\mathbf{z}$ is the only point in $L + \mathbf{z}$ with ℓ_1-norm at most 1. That is, $(L + \mathbf{z}) \cap B_1^n = \{\mathbf{z}\}$ as claimed.

Conversely, we assume that L satisfies the condition in the lemma and we consider $\mathbf{b} \in \mathbb{R}^m$. Let us suppose that the system $A\mathbf{x} = \mathbf{b}$ has a solution $\tilde{\mathbf{x}}$ with at most r nonzero components. If $\tilde{\mathbf{x}} = \mathbf{0}$, then $\mathbf{b} = \mathbf{0}$, and clearly, $\mathbf{0}$ is also the only optimum of (BP). For $\tilde{\mathbf{x}} \neq \mathbf{0}$, we set $\mathbf{z} = \frac{\tilde{\mathbf{x}}}{\|\tilde{\mathbf{x}}\|_1}$. Then $\|\mathbf{z}\|_1 = 1$ and $|\mathrm{supp}(\mathbf{z})| \leq r$, and so by the assumption, $\mathbf{z}$ is the only point in $L + \mathbf{z}$ of ℓ_1-norm at most 1. By rescaling we get that $\tilde{\mathbf{x}}$ is the only point in $L + \tilde{\mathbf{x}}$ of ℓ_1-norm at most $\|\tilde{\mathbf{x}}\|_1$, and since $L + \tilde{\mathbf{x}}$ is the set of all solutions of $A\mathbf{x} = \mathbf{b}$, we get that A is BP-exact. $\qquad\square$

Intuition for BP-exactness. We don't have means for proving Theorem 8.5.2, but now, using the lemma just proved, we can at least try to convey some intuition as to why a claim like Theorem 8.5.2 is plausible, and what kind of calculations are needed to prove it.

The kernel L of a random matrix A defines a random k-dimensional subspace[6] of $\mathbb{R}^n$, where $k = n-m$. For proving Theorem 8.5.2, we need

---

[6] The question, "What is a random k-dimensional subspace?" is a subtle one. For us, the simplest way out is to define a random k-dimensional subspace as the

to verify that L is *good* for every boundary point $\mathbf{z}$ of the crosspolytope with $|\mathrm{supp}(\mathbf{z})| \leq r$, where we say that L is good for $\mathbf{z}$ if $(L+\mathbf{z}) \cap B_1^n = \{\mathbf{z}\}$.

For $\mathbf{z}$ as above, let us define a convex cone $C_{\mathbf{z}} = \{t(\mathbf{x} - \mathbf{z}) : t \geq 0, \ \mathbf{x} \in B_1^n\}$. Geometrically, we take the cone generated by all rays emanating from $\mathbf{z}$ and intersecting the crosspolytope in a point other than $\mathbf{z}$, and we translate the cone so that $\mathbf{z}$ is moved to the origin. Then L good for $\mathbf{z}$ means exactly that $L \cap C_{\mathbf{z}} = \{\mathbf{0}\}$.

The points $\mathbf{z}$ on the boundary with at most r nonzero coordinates fill out exactly the union of all $(r-1)$-dimensional faces of the crosspolytope. Let F be one of these faces. It can be checked that the cone $C_{\mathbf{z}}$ is the same for all $\mathbf{z}$ in the relative interior of F, so we can define the cone C_F associated with the face F (the reader may want to consider some examples for the 3-dimensional regular octahedron). Moreover, if $\mathbf{y}$ is a boundary point of F, then $C_{\mathbf{y}} \subseteq C_F$, and so if L is good for some point in the relative interior of F, then it is good for all points of F including the boundary.

Let p_F denote the probability that a random L is *bad* (i.e., not good) for some $\mathbf{z} \in F$. Then the probability that L is bad for any $\mathbf{z}$ at all is no more than $\sum_F p_F$, where the sum is over all $(r-1)$-dimensional faces of the crosspolytope.

It is not too difficult to see that the number of $(r-1)$-dimensional faces is $\binom{n}{r} 2^r$, and that the cones C_F of all of these faces are congruent (they differ only by rotation around the origin). Therefore, all p_F equal the same number $p = p(n, k, r)$, and the probability of L bad for at least one $\mathbf{z}$ is at most $\binom{n}{r} 2^r p$. If we manage to show, for some values of n, k, and r, that the expression $\binom{n}{r} 2^r p$ is much smaller than 1, then we can conclude that a random matrix A is BP-exact with high probability.

Estimating $p(n, k, r)$ is a nontrivial task; its difficulty heavily depends on the accuracy we want to attain. Getting an estimate that is more or less accurate including numerical constants, such as is needed to prove Theorem 8.5.2, is quite challenging. On the other hand, if we don't care about numerical constants so much and want just a rough asymptotic result, standard methods from high-dimensional convexity theory lead to the goal much faster.

---

kernel of a random $m \times n$ matrix with independent normal entries as in Theorem 8.5.2. Fortunately, this turns out to be equivalent to the usual (and "right") definition, which is as follows. One fixes a particular k-dimensional subspace R_0, say the span of the first k vectors of the standard basis, and defines a random subspace as a random rotation of R_0. This may not sound like great progress, since we have just used the equally problematic-looking notion of random rotation. But the group $SO(n)$ of all rotations in $\mathbb{R}^n$ around the origin is a compact group and hence it has a unique invariant probability measure (Haar measure), which defines "random rotation" satisfactorily and uniquely.

Here we conclude this very rough outline of the argument with a few words on why one should expect $p(n, k, r)$ to be very small for k and r much smaller than n. Roughly speaking, this is because for n large, the n-dimensional crosspolytope is a very lean and spiky body, and the cones C_F are very narrow for low-dimensional faces F. Hence a random subspace L of not too large dimension is very likely to avoid C_F. As a very simplified example of this phenomenon, we may consider $k = r = 1$. Then F is a vertex and C_F is easily described. As a manageable exercise, the reader may try to estimate the fraction of the unit sphere centered at $\mathbf{0}$ that is covered by C_F; this quantity is exactly half of the probability $p(n, 1, 1)$ that a random line through $\mathbf{0}$ intersects C_F nontrivially.

References. Basis pursuit was introduced in

S. Chen, D. L. Donoho, and M. A. Saunders: Atomic decomposition by basis pursuit, *SIAM J. Scientific Computing* 20, 1(1999) 33–61.

A classical approach to finding a "good" solution of an underdetermined system $A\mathbf{x} = \mathbf{b}$ would be to minimize $\|\mathbf{x}\|_2$, rather than $\|\mathbf{x}\|_1$ (a "least squares" or "generalized inverse" method), which typically yields a solution with many nonzero components and is much less successful in applications such as the decoding problem. Basis pursuit, by minimizing the ℓ_1-norm instead, yields a basic solution with only a few nonzero components.

Interestingly, several groups of researchers independently arrived at the concept of BP-exactness and obtained the following general version of Theorem 8.5.2: *For every constant $\alpha \in (0, 1)$ there exists $\beta = \beta(\alpha) > 0$ such that a random $\lfloor \alpha n \rfloor \times n$ matrix A is BP-exact for $r = \lfloor \beta n \rfloor$ with probability exponentially close to 1.* Combined results of two of these groups can be found in

E. J. Candès, M. Rudelson, T. Tao, and R. Vershynin: Error correction via linear programming, *Proc. 46th IEEE Symposium on Foundations of Computer Science* (FOCS), 2005, pages 295–308.

A third independent proof was given by Donoho. Later, and by yet another method, he obtained the strongest known quantitative bounds, including those in Theorem 8.5.2 (the previously mentioned proofs yield a much smaller constant for $\alpha = 0.75$ than 0.08). Among his several papers on the subject we cite

D. Donoho: High-dimensional centrally symmetric polytopes with neighborliness proportional to dimension, *Discrete and Computational Geometry* 35(2006), 617–652, where he establishes a connection to the classical theory of convex polytopes using a result in the spirit of Lemma 8.5.4 and the remarks following it (but more elaborate). Through this connection he obtained an interesting *upper* bound on $\beta(\alpha)$. For example, in the setting of Theorem 8.5.2, there exists no $\lfloor 0.75n \rfloor \times n$ BP-exact matrix at all with $r > 0.25n$ (assuming n large). Additional upper bounds, essentially showing the existence results for BP-matrices in the above papers to be asymptotically optimal, were proved by

> N. Linial and I. Novik: How neighborly can a centrally symmetric polytope be?, *Discr. Comput. Geom.*, in press.

We remark that our notation is a compromise among the notations of the papers quoted above and doesn't follow any of them exactly, and that the term "BP-exact" is ours.

More applications of sparse solutions of underdetermined systems. The problem of computing a sparse solution of a system of linear equations arises in *signal processing*. The signal considered may be a recording of a sound, a measurement of seismic waves, a picture taken by a digital camera, or any of a number of other things. A classical method of analyzing signals is *Fourier analysis*, which from a linear-algebraic point of view means expressing a given periodic function in a basis consisting of the functions $1, \cos x, \sin x, \cos 2x, \sin 2x, \cos 3x, \sin 3x, \ldots$ (the closely related *cosine transform* is used in the *JPEG encoding* of digital pictures). These functions are linearly independent, and so the expression (Fourier series) is unique. In the more recent *wavelet* analysis[7] one typically has a larger collection of basic functions, the wavelets, which can be of various kinds, depending on the particular application. They are no longer linearly independent, and hence there are many different ways of representing a given signal as a linear combination of wavelets. So one looks for a representation satisfying some additional criteria, and sparsity (small number of nonzero coefficients) is a very natural criterion: It leads to an economic (compressed) representation, and sometimes it may also help in analyzing or filtering the signal. For example, let us imagine that there is a smooth signal that has a nice representation by sine and cosine functions, and then an impulsive noise made of "spike" functions is added to it. We let the basic functions be sines and cosines *and* suitable spike functions, and by computing a sparse representation in such a basis we can often isolate the noise component very well, something that the classical Fourier analysis cannot do. Thus, we naturally arrive at computing sparse solutions of underdetermined linear systems.

Another source is computer tomography (CT), where one has an unknown vector $\mathbf{x}$ (each x_i is the density of some small area of tissue,

---

[7] Indeed, the newer picture encoding standard JPEG 2000 employs wavelets.

say), and the CT scanner measures various linear combinations of the x_i, corresponding to rays through the tissue in various directions. Sometimes there are reasons to expect that only a small number of the pixels will have values x_i different from the background level, and when we want to reconstruct $\mathbf{x}$ from the scan, we again ask for a sparse solution of a linear system. (More realistically, although less intuitively, we don't expect a small number of nonzero *pixels*, but rather a small number of significantly nonzero coefficients in a suitable wavelet representation.)

8.6 Transversals of d-Intervals

This section describes an application of the duality theorem in discrete geometry and combinatorics. We begin with a particular geometric result. Then we discuss concepts appearing in the proof in a more general context.

Helly's and Gallai's theorems. First let $\mathcal{I} = \{I_1, I_2, \ldots, I_n\}$ be a family of closed intervals on the real line such that every two of the intervals intersect. It is easily seen that there exists a point common to all of the intervals in $\mathcal{I}$: Indeed, the rightmost among the left endpoints of the I_i is such a point. In more detail, writing $I_i = [a_i, b_i]$ and setting $a = \max\{a_1, \ldots, a_n\}$, we necessarily have $a_i \leq a \leq b_i$ for all i, since $a_i \leq a$ is immediate from the definition of a, and if we had $a > b_i$ for some i, then $I_i = [a_i, b_i]$ would be disjoint from the interval beginning with a.

The statement just proved is a special (one-dimensional) case of a beautiful and important theorem of Helly: *If $C_1, C_2, \ldots, C_n$ are convex sets in $\mathbb{R}^d$ such that any at most $d + 1$ of them have a point in common, then there is a point common to all of the C_i.* We will not prove this result; it is mentioned as a background against which the forthcoming results can be better appreciated.

It is easily seen that in general we cannot replace $d + 1$ by any smaller number in Helly's theorem. For example, in the plane, the assumption of Helly's theorem requires every three of the sets to have a common point, and pairwise intersections are not enough. To see this, we consider n lines in general position. They are convex sets, every two of them intersect, but no three have a common point.

Let us now consider planar convex sets of a special kind, namely, circular disks. One can easily draw three disks such that every two intersect, but no point is common to all three. However, there is a theorem (Gallai's) for pairwise intersecting disks in the spirit of Helly's theorem: *If $\mathcal{D} = \{D_1, D_2, \ldots, D_n\}$ is a family of disks in the plane in which every two disks intersect, then there exist 4 points such that each D_i contains at least* *one of them.* With the (best possible) constant 4 this is a quite difficult theorem, but it is not too hard to prove a similar theorem with 4 replaced by some large constant. The reader is invited to solve this as a puzzle.

A set of points as in the theorem that intersects every member of $\mathcal{D}$ is called a **transversal** of $\mathcal{D}$ (sometimes one also speaks of *piercing* or *stabbing* $\mathcal{D}$ by a small number of points). Thus pairwise intersecting disks in the plane always have a 4-point transversal, and Helly's theorem asserts that $(d+1)$-wise intersecting convex sets in $\mathbb{R}^d$ have a one-point transversal.

What conditions on a family of sets guarantee that it has a small transversal? This fairly general question subsumes many interesting particular problems and it has been the subject of much research. Here we consider a one-dimensional situation. At the beginning of the section we dealt with a family of intervals, and now we admit intervals with some bounded number of "holes."

Transversals for pairwise intersecting d-intervals.   For an integer $d \geq 1$, a **d-interval** is defined as the union of d closed intervals on the real line. The following picture shows three pairwise intersecting 2-intervals (drawn solid, dashed, and dash-dotted, respectively) with no point common to all three:

Thus, we cannot expect a one-point transversal for pairwise intersecting d-intervals. But the following theorem shows the existence of a transversal whose size depends only on d:

8.6.1 Theorem. *Let $\mathcal{J}$ be a finite family of d-intervals such that $J_1 \cap J_2 \neq \emptyset$ for every $J_1, J_2 \in \mathcal{J}$. Then $\mathcal{J}$ has a transversal of size $2d^2$; that is, there exist $2d^2$ points such that each d-interval of $\mathcal{J}$ contains at least one of them.*

At first sight it is not obvious that there is any bound at all for the size of the transversal that depends only on d. This was first proved in 1970, with a bound exponential in d, in

A. Gyárfás, J. Lehel: A Helly-type problem in trees, in *Combinatorial Theory and its Applications*, P. Erdős, A. Rényi, and V.T. Sós, editors, North-Holland, Amsterdam, 1970, pages 571–584.

The best bound known at present is d^2, and it has been established using algebraic topology in

T. Kaiser: Transversals of d-intervals, *Discrete Comput. Geom.* 18(1997) 195–203.

We are going to prove a bound that is worse by a factor of 2, but the proof presented here is much simpler. It comes from

N. Alon: Piercing d-intervals, *Discrete Comput. Geom.* 19(1998) 333–334,

following a general method developed in

N. Alon, D. Kleitman: Piercing convex sets and the Hadwiger–Debrunner (p, q)-problem, *Adv. Math.* 96(1992) 103–112.

It is also known that the bound cannot be improved below a constant multiple of $\frac{d^2}{\log d}$; see

J. Matoušek: Lower bounds on the transversal numbers of d-intervals, *Discrete Comput. Geom.* 26(2001) 283–287.

Working toward a proof of the theorem, we first show that in a family of pairwise intersecting d-intervals, some point has to be contained in "many" members of the family. For reasons that will become apparent later, we formulate it for a finite *sequence* of d-intervals, so that repetitions are allowed.

8.6.2 Lemma. *Let $J_1, J_2, \ldots, J_n$ be d-intervals such that $J_i \cap J_j \neq \emptyset$ for all $i, j \in \{1, 2, \ldots, n\}$. Then there is an endpoint of some J_i that is contained in at least $n/2d$ of the d-intervals.*

Proof. Let T be the set of all ordered triples (p, i, j) such that p is one of the at most d *left* endpoints of J_i, and $p \in J_j$. We want to bound the size of T from below. Let us fix $i \leq j$ for a moment. Let p be the leftmost point of the (nonempty) intersection $J_i \cap J_j$. Clearly, p is a left endpoint of J_i or J_j, and thus T contains one of the triples (p, i, j) or (p, j, i). So every pair i, j, $i \leq j$, contributes at least one member of T, and thus $|T| \geq \binom{n}{2} + n \geq n^2/2$. Since there are at most dn possible values for the pair (p, i), it follows that there exist p_0 and i_0 such that $(p_0, i_0, j) \in T$ for at least $(n^2/2)/dn = n/2d$ values of j. This means that p_0 is contained in at least $n/2d$ of the J_j.   □

Next, we show that for every family of pairwise intersecting d-intervals we can distribute weights on the endpoints such that every d-interval has relatively large weight (compared to the total weight of all points).

8.6.3 Lemma. *Let $\mathcal{J}$ be a finite family of pairwise intersecting d-intervals, and let P denote the set of endpoints of the d-intervals in $\mathcal{J}$. Then there are nonnegative real numbers x_p, $p \in P$, such that*

$$\sum_{p \in J \cap P} x_p \geq 1 \text{ for every } J \in \mathcal{J}, \text{ and}$$

$$\sum_{p \in P} x_p \leq 2d.$$

Before proving the lemma, let us see how it implies Theorem 8.6.1. Given $\mathcal{J}$ and the weights as in the lemma, we will choose a set $X \subseteq P$ such that $|X| \leq 2d^2$, and each of the $|X| + 1$ open intervals into which X divides the real axis contains points of P of total weight less than $1/d$. Such an X can be selected from P by a simple left-to-right scan: Let $p_1 < p_2 < \cdots < p_m$ be the points of P. We consider them one by one in this order, and we include p_i into X whenever $\sum_{j:\, p_\ell < p_j \leq p_i} x_{p_j} \geq \frac{1}{d}$, where p_ℓ is the last point already included in X (and we formally put $p_\ell = -\infty$ if no point has yet been included in X). It is clear that none of the open intervals determined by X contains weight $1/d$ or larger, and the bound on the size of X follows easily, using that the total weight of all points of P is at most $2d$.

We claim that X is a transversal of $\mathcal{J}$. Indeed, considering a $J \in \mathcal{J}$, at least one of the d components of J is an interval containing points of P of total weight at least $1/d$, and so it contains a point of X.

Proof of Lemma 8.6.3 by duality. We formulate the problem of choosing the weights x_p as a linear program with variables x_p, $p \in P$:

$$
\begin{aligned}
&\text{Minimize} && \textstyle\sum_{p \in P} x_p \\
&\text{subject to} && \textstyle\sum_{p \in J \cap P} x_p \geq 1 \text{ for every } J \in \mathcal{J}, \\
& && \mathbf{x} \geq \mathbf{0}.
\end{aligned}
$$

The linear program is certainly feasible; for instance, $x_p = 1$ for all p is a feasible solution. We would like to show that the optimum is at most $2d$.

Using the dualization recipe from Section 6.2, we find that the dual linear program has variables y_J, $J \in \mathcal{J}$, and it looks as follows:

$$
\begin{aligned}
&\text{Maximize} && \textstyle\sum_{J \in \mathcal{J}} y_J \\
&\text{subject to} && \textstyle\sum_{J:\, p \in J \cap P} y_J \leq 1 \text{ for every } p \in P, \\
& && \mathbf{y} \geq \mathbf{0}.
\end{aligned}
$$

The dual linear program is feasible, too, since $\mathbf{y} = \mathbf{0}$ is feasible. Then by the duality theorem both the primal and the dual linear programs have optimal solutions, $\mathbf{x}^*$ and $\mathbf{y}^*$, respectively, that yield the same value of the objective functions.

We may assume that $\mathbf{y}^*$ is rational: Indeed, we may take it to be a basic feasible solution, and all basic feasible solutions are rational since all coefficients of the linear program are rational.

We now have some rational weight y_J^* for every $J \in \mathcal{J}$ such that no point of P is contained in d-intervals of total weight exceeding 1, and we want to show that the total weight of all $J \in \mathcal{J}$ cannot be larger than $2d$.

Lemma 8.6.2 tells us that if all the d-intervals had the same weight, then there would be a point contained in d-intervals of weight at least $W/2d$, where W is the total weight of all d-intervals. Our weights need not all be equal, but we will pass to the case of equal weights by replicating each d-interval a suitable number of times.

Let D be a common denominator of all the rational numbers y_J^*, $J \in \mathcal{J}$, and let $y_J^* = \frac{r_J}{D}$, with r_J integral. Let $(J_1, J_2, \ldots, J_n)$ be a sequence that includes each d-interval $J \in \mathcal{J}$ exactly r_J times (thus $n = \sum_{J \in \mathcal{J}} r_J$). By Lemma 8.6.2 there is a point $p \in P$ contained in at least $n/2d$ members of the sequence, which means that

$$\sum_{J \in \mathcal{J}: p \in J} r_J \geq \frac{n}{2d} = \frac{1}{2d} \sum_{J \in \mathcal{J}} r_J.$$

Dividing both sides by the common denominator D and multiplying by $2d$ gives

$$2d \cdot \sum_{J \in \mathcal{J}: p \in J} y_J^* \geq \sum_{J \in \mathcal{J}} y_J^*.$$

Since $\mathbf{y}^*$ is a feasible solution of the dual linear program, the left-hand side is at most $2d$, and so $\sum_{J \in \mathcal{J}} y_J^* \leq 2d$. This concludes the proof of Lemma 8.6.3 as well as of Theorem 8.6.1. □

Transversal number and matching number. Let us now look at some of the concepts appearing in the proof of Theorem 8.6.1 in a general context. Let V be an arbitrary finite set, and let $\mathcal{F}$ be a system of subsets of V.

A set $X \subseteq V$ is called a *transversal* of $\mathcal{F}$ if $F \cap X \neq \emptyset$ for every $F \in \mathcal{F}$. The **transversal number** $\tau(\mathcal{F})$ is the minimum possible number of elements of a transversal of $\mathcal{F}$.

Determining or estimating the transversal number of a given set system is an important basic problem in combinatorics and combinatorial optimization, including many other problems as special cases. For example, if we consider a graph $G = (V, E)$, and view the edges as two-element subsets of V, then a transversal of E is exactly what was called a vertex cover in Section 3.3.

Another useful notion is the **matching number** of $\mathcal{F}$, denoted by $\nu(\mathcal{F})$ and defined as the maximum number of sets in a subsystem $\mathcal{M} \subseteq \mathcal{F}$ such that no two distinct sets $F_1, F_2 \in \mathcal{M}$ intersect (such an $\mathcal{M}$ is called a *matching*).

It is easily seen that $\nu(\mathcal{F}) \leq \tau(\mathcal{F})$ for every $\mathcal{F}$: If $\mathcal{F}$ has matching number k, then $\mathcal{F}$ contains k pairwise disjoint sets; thus, we need at least k points to get a transversal of $\mathcal{F}$.

For the graph example, $\nu(E)$ is exactly the number of edges in a maximum matching. This is also where the name "matching number" comes from.

The condition in Theorem 8.6.1 that every two d-intervals in $\mathcal{J}$ intersect can be rephrased as $\nu(\mathcal{J}) = 1$. More generally, $\nu(\mathcal{F}) \leq k$ means that among every $k + 1$ members of $\mathcal{F}$ we can find two that intersect. There is a more general version of Theorem 8.6.1 stating that $\tau(\mathcal{J}) \leq 2d^2\nu(\mathcal{J})$ for every finite family of d-intervals. The proof is very similar to the one shown for Theorem 8.6.1, except that the analogue of Lemma 8.6.2 needs the well-known Turán's theorem from graph theory.

Fractional transversals and matchings. In the proof of Theorem 8.6.1 we have implicitly used another parameter of a set system, which always lies between $\nu(\mathcal{F})$ and $\tau(\mathcal{F})$ and which, unlike $\tau(\mathcal{F})$ and $\nu(\mathcal{F})$, is efficiently computable. This new parameter can be introduced in two seemingly different ways, which turn out to be equivalent by the duality theorem of linear programming.

A **fractional transversal** of $\mathcal{F}$ is any feasible solution $\mathbf{x}$ to the linear program

$$\begin{array}{ll} \text{minimize} & \sum_{v \in V} x_v \\ \text{subject to} & \sum_{v \in F} x_v \geq 1 \text{ for every } F \in \mathcal{F}, \\ & \mathbf{x} \geq \mathbf{0}. \end{array}$$

So in a fractional transversal one can take, say, one-third of one point and two-thirds of another, but for each set, the fractions for points in that set must add up to at least 1, one full point. The **fractional transversal number** $\tau^*(\mathcal{F})$ is the optimal value of the objective function, i.e., the minimum possible total weight of a fractional transversal.

Every transversal T corresponds to a fractional transversal, given by $x_v = 1$ if $v \in T$ and $x_v = 0$ otherwise, and thus $\tau^*(\mathcal{F}) \leq \tau(\mathcal{F})$ for every $\mathcal{F}$.

A **fractional matching** for $\mathcal{F}$ is any feasible solution $\mathbf{y}$ to the linear program

$$\begin{array}{ll} \text{maximize} & \sum y_F \\ \text{subject to} & \sum_{F: v \in F} y_F \leq 1 \text{ for every } v \in V, \\ & \mathbf{y} \geq \mathbf{0}, \end{array}$$

and the optimal value of the objective function is the *fractional matching number* $\nu^*(\mathcal{F})$.

Every matching $\mathcal{M}$ yields a fractional matching (we put $y_F = 1$ for $F \in \mathcal{M}$ and $y_F = 0$ otherwise). Thus, $\nu(\mathcal{F}) \leq \nu^*(\mathcal{F})$.

Since the linear programs for τ^* and for ν^* are dual to each other, we always have $\nu^*(\mathcal{F}) = \tau^*(\mathcal{F})$, and altogether we have the chain of inequalities

$$\nu(\mathcal{F}) \leq \nu^*(\mathcal{F}) = \tau^*(\mathcal{F}) \leq \tau(\mathcal{F}).$$

We remark that if $\mathcal{F}$ is the set of edges of a bipartite graph, then König's theorem (Theorem 8.2.2) asserts exactly that $\nu(\mathcal{F}) = \tau(\mathcal{F})$. On the other hand, if $\mathcal{F}$ is the edge set of a triangle (that is, $\mathcal{F} = \{\{1,2\}, \{1,3\}, \{2,3\}\}$), then $\nu(\mathcal{F}) = 1 < \nu^*(\mathcal{F}) = \frac{3}{2} < \tau(\mathcal{F}) = 2$.

The proof of Theorem 8.6.1 can now be viewed as follows: First one proves that $\nu^*(\mathcal{J}) \leq 2d$ for every family of d-intervals with $\nu(\mathcal{J}) = 1$, and then one shows that $\tau(\mathcal{J}) \leq d \cdot \tau^*(\mathcal{J})$. This proof scheme turned out to be very powerful and it works in many other cases as well.

Fractional concepts. Besides the fractional matching and transversal numbers, many other "fractional" quantities appear in combinatorics and combinatorial optimization. The general recipe is to take some useful integer-valued parameter Q of a graph, say, reformulate its definition as an integer program, and introduce a "fractional Q" as the optimum value of a suitable LP relaxation of the integer program. In many cases such a fractional Q is useful for studying or approximating the original Q. The book

> E. R. Scheinerman and D. H. Ullman: *Fractional Graph Theory*, John Wiley & Sons, New York 1997,

nicely presents such developments. Let us conclude this section by quoting an example from that book. We consider five committees, numbered 1,2,...,5, such that 1 and 2 have a common member, and so have 2 and 3, 3 and 4, 4 and 5, and 5 and 1, while any other pair of committees is disjoint. A one-hour meeting should be scheduled for each committee, and meetings of committees with a common member must not overlap. What is the length of the shortest time interval in which all the five meetings can be scheduled?

It seems that a 3-hour schedule like the one below should be optimal:

12:00   13:00   14:00   15:00

1	2	5
3	4	

However, if one of the committees is willing to break its meeting into two half-hour parts, then a shorter schedule is possible:

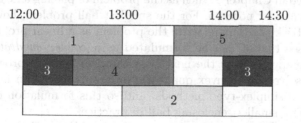

The first schedule corresponds to a proper coloring of the conflict graph

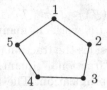

by three colors, while the second schedule corresponds to a *fractional coloring* of the same graph, with value 2.5.

8.7 Smallest Balls and Convex Programming

The smallest ball problem. We are given points $\mathbf{p}_1, \ldots, \mathbf{p}_n \in \mathbb{R}^d$, and we want to find a ball of the smallest radius that contains all the points.[8]

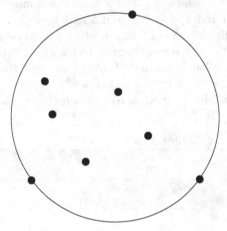

This looks similar to some of the geometric optimization problems that we have addressed in Chapter 2, such as the problem of placing a largest possible disk inside a convex polygon. For the smallest ball problem, however, there is no simple trick that lets us write the problem as a linear program.

We will see that it can be formulated as a *convex quadratic program*, which is in many respects the next best thing to a linear program. There are efficient solvers for convex quadratic programs, based on interior point methods or on simplex-type methods, and so this formulation can be used for computing a smallest enclosing ball in practice.

We will also derive from this formulation that the smallest enclosing ball always exists, and it is determined uniquely. (This is intuitively very plausible; think of a shrinking rubber ball.) In the course of the proof, we will establish

---

[8] In the plane this is sometimes referred to as the *smallest bomb problem*, but we prefer not to elaborate this association into a real-life story.

a powerful criterion for optimality of a feasible solution of a convex program, known (in a much more general context) as the *Karush–Kuhn–Tucker conditions*. These conditions are of outstanding theoretical value, and they are the basis of efficient solution methods for many classes of optimization problems, including convex quadratic programming. The reader might still wonder, how does all of this relate to linear programming? We will use the duality theorem of linear programming to derive the Karush–Kuhn–Tucker conditions.

We begin by introducing convex programming, and we will return to the smallest enclosing ball problem later.

A short introduction to convex programming. Let us recall that an n-variate function $f\colon \mathbb{R}^n \to \mathbb{R}$ is *convex* if

$$f\big((1-t)\mathbf{x} + t\mathbf{y}\big) \le (1-t)f(\mathbf{x}) + tf(\mathbf{y})$$

for all $\mathbf{x}, \mathbf{y} \in \mathbb{R}^n$ and all $t \in [0,1]$. Geometrically, the segment connecting the points $(\mathbf{x}, f(\mathbf{x}))$ and $(\mathbf{y}, f(\mathbf{y}))$ in $\mathbb{R}^{n+1}$ never goes below the graph of f.

A **convex program** is the problem of minimizing a convex function in n variables subject to linear equality and inequality constraints.[9]

The following picture illustrates a 2-dimensional convex programming problem, with a planar feasible region given by four inequality constraints:

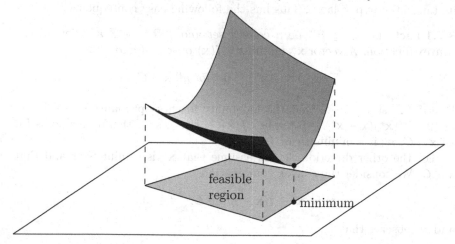

We note that the minimum need not occur at a vertex of the feasible region. Moreover, an optimal solution need not exist even if the convex function $f(\mathbf{x})$ is bounded from below; an example is the problem of minimizing e^{-x} subject to $x \ge 0$. We should also remark that, as is possible in linear programming, we cannot change minimization to maximization, since for f convex, $-f$ is typically not convex (unless f is linear). Actually, *maximizing* a convex function subject to linear constraints is a computationally difficult (NP-hard) problem.

---

[9] Some sources allow other types of convex constraints in a convex program, but we don't need this here.

Here we will consider convex programs in equational form:

$$\begin{array}{ll}\text{Minimize} & f(\mathbf{x}) \\ \text{subject to} & A\mathbf{x} = \mathbf{b} \\ & \mathbf{x} \geq 0, \end{array} \qquad (8.11)$$

where $A \in \mathbb{R}^{m \times n}$, $\mathbf{b} \in \mathbb{R}^m$, and $f: \mathbb{R}^n \to \mathbb{R}$ is a convex function.

In order to use calculus, we also assume that f is differentiable, with continuous partial derivatives. In this situation, the inequality

$$f(\mathbf{x}) \geq f(\mathbf{z}) + \nabla f(\mathbf{z})(\mathbf{x} - \mathbf{z}) \qquad (8.12)$$

holds for all $\mathbf{x}, \mathbf{z} \in \mathbb{R}^n$, and this is an alternative characterization of convexity. We recall that

$$\nabla f(\mathbf{z}) = \left(\frac{\partial}{\partial x_1} f(\mathbf{x})|_{\mathbf{x}=\mathbf{z}}, \dots, \frac{\partial}{\partial x_n} f(\mathbf{x})|_{\mathbf{x}=\mathbf{z}} \right)$$

is the gradient (vector of partial derivatives) of f at $\mathbf{z}$. Thus $\nabla f(\mathbf{z})(\mathbf{x} - \mathbf{z})$ is the scalar product of the row vector $\nabla f(\mathbf{z})$ with the column vector $\mathbf{x} - \mathbf{z}$.

Geometrically, the inequality says that the epigraph of f lies above all of its tangential hyperplanes. This has the following easy consequence.

8.7.1 Fact. *Let $C \subseteq \mathbb{R}^n$ be a convex set and $f: \mathbb{R}^n \to \mathbb{R}$ a differentiable convex function. A vector $\mathbf{x}^*$ minimizes $f(\mathbf{x})$ over C if and only if*

$$\nabla f(\mathbf{x}^*)(\mathbf{x} - \mathbf{x}^*) \geq 0 \quad \text{for all } \mathbf{x} \in C.$$

Proof. First we prove that the inequality implies optimality of $\mathbf{x}^*$. Using (8.12), $\nabla f(\mathbf{x}^*)(\mathbf{x} - \mathbf{x}^*) \geq 0$ implies $f(\mathbf{x}) \geq f(\mathbf{x}^*)$, so if the former holds for all $\mathbf{x} \in C$, $\mathbf{x}^*$ is a minimizer of f over C.

For the other direction, let us assume that $\mathbf{x}^*$ is a minimizer and that $\mathbf{x} \in C$. We consider the convex combination

$$\mathbf{x}(t) := \mathbf{x}^* + t(\mathbf{x} - \mathbf{x}^*) \in C, \quad t \in [0, 1],$$

and we observe that

$$\frac{\partial}{\partial t} f(\mathbf{x}(t))|_{t=0} = \lim_{t \to 0} \frac{f(\mathbf{x}(t)) - f(\mathbf{x}^*)}{t} \geq 0$$

must hold, for otherwise, $f(\mathbf{x}(t)) < f(\mathbf{x}^*)$ for some small t. On the other hand, we have

$$\frac{\partial}{\partial t} f(\mathbf{x}(t))|_{t=0} = \nabla f(\mathbf{x}^*)(\mathbf{x} - \mathbf{x}^*),$$

by the chain rule. This completes the proof. $\qquad \square$

Next we formulate and prove the promised optimality criterion for convex programming.

8.7.2 Proposition (Karush–Kuhn–Tucker conditions). *Let us consider the convex program*

$$\begin{aligned} \text{minimize} \quad & f(\mathbf{x}) \\ \text{subject to} \quad & A\mathbf{x} = \mathbf{b} \\ & \mathbf{x} \geq \mathbf{0} \end{aligned}$$

with f convex and differentiable, with continuous partial derivatives. A feasible solution $\mathbf{x}^* \in \mathbb{R}^n$ is optimal if and only if there is a vector $\tilde{\mathbf{y}} \in \mathbb{R}^m$ such that for all $j \in \{1, \ldots, n\}$,

$$\nabla f(\mathbf{x}^*)_j + \tilde{\mathbf{y}}^T \mathbf{a}_j \begin{cases} = 0 & \text{if } x_j^* > 0 \\ \geq 0 & \text{otherwise.} \end{cases}$$

Here $\mathbf{a}_j$ is the jth column of A.

*The components of $\tilde{\mathbf{y}}$ are called the **Karush–Kuhn–Tucker multipliers**.*

Proof. First we assume that there is a vector $\tilde{\mathbf{y}}$ with the above properties, and we let $\mathbf{x}$ be any feasible solution to the convex program. Then we get

$$\begin{aligned} \left(\nabla f(\mathbf{x}^*) + \tilde{\mathbf{y}}^T A \right) \mathbf{x}^* &= 0, \\ \left(\nabla f(\mathbf{x}^*) + \tilde{\mathbf{y}}^T A \right) \mathbf{x} &\geq 0. \end{aligned}$$

Subtracting the first equation from the second, the contributions of $\tilde{\mathbf{y}}^T A$ cancel (since $A\mathbf{x} = \mathbf{b} = A\mathbf{x}^*$ by the feasibility of $\mathbf{x}$ and $\mathbf{x}^*$), and we conclude that

$$\nabla f(\mathbf{x}^*)(\mathbf{x} - \mathbf{x}^*) \geq 0.$$

Since this holds for all feasible solutions $\mathbf{x}$, the solution $\mathbf{x}^*$ is optimal by Fact 8.7.1.

Conversely, let $\mathbf{x}^*$ be optimal, and let us set $\mathbf{c}^T = -\nabla f(\mathbf{x}^*)$. By Fact 8.7.1 we have $\mathbf{c}^T(\mathbf{x} - \mathbf{x}^*) \leq 0$ for all feasible solutions $\mathbf{x}$, meaning that $\mathbf{x}^*$ is an optimal solution of the linear program

$$\begin{aligned} \text{maximize} \quad & \mathbf{c}^T \mathbf{x} \\ \text{subject to} \quad & A\mathbf{x} = \mathbf{b} \\ & \mathbf{x} \geq \mathbf{0}. \end{aligned}$$

According to the dualization recipe (Section 6.2), its dual is the linear program

$$\begin{aligned} \text{minimize} \quad & \mathbf{b}^T \mathbf{y} \\ \text{subject to} \quad & A^T \mathbf{y} \geq \mathbf{c}, \end{aligned}$$

and the duality theorem implies that it has an optimal solution $\tilde{\mathbf{y}}$ satisfying

$$\mathbf{b}^T \tilde{\mathbf{y}} = \mathbf{c}^T \mathbf{x}^*. \tag{8.13}$$

Since $\tilde{\mathbf{y}}$ is a feasible solution of the dual linear program, we have $\tilde{\mathbf{y}}^T\mathbf{a}_j \geq c_j$ for all j, and (8.13) implies

$$(\tilde{\mathbf{y}}^T A - \mathbf{c}^T)\mathbf{x}^* = \mathbf{b}^T\tilde{\mathbf{y}} - \mathbf{c}^T\mathbf{x}^* = 0.$$

So we have $\nabla f(\mathbf{x}^*)_j + \tilde{\mathbf{y}}^T\mathbf{a}_j = -c_j + \tilde{\mathbf{y}}^T\mathbf{a}_j \geq 0$, with equality whenever $x_j^* > 0$. Therefore, we have found the desired multipliers $\tilde{\mathbf{y}}$.   □

The fact that there is dualization for everyone implies Karush–Kuhn–Tucker conditions for everyone. We encourage the reader to work out the details, and we mention only one special case here: A feasible solution $\mathbf{x}^*$ of the convex programming problem

$$\begin{array}{ll} \text{minimize} & f(\mathbf{x}) \\ \text{subject to} & A\mathbf{x} = \mathbf{b} \end{array}$$

is optimal if and only if there exists a vector $\tilde{\mathbf{y}}$ such that

$$\nabla f(\mathbf{x}^*) + \tilde{\mathbf{y}}^T A = \mathbf{0}^T.$$

In this special case, the components of $\tilde{\mathbf{y}}$ are called **Lagrange multipliers** and can be obtained from $\mathbf{x}^*$ through Gaussian elimination (also see Section 7.2, where the method of Lagrange multipliers is briefly described in a more general setting). If, in addition, f is a quadratic function, its gradient is linear, so the minimization problem itself (finding an optimal $\mathbf{x}^*$ with a matching $\mathbf{y}$) can be solved through Gaussian elimination. For example, the problem of fitting a line by the method of *least squares*, mentioned in Section 2.4, is of this easy type, because its bivariate quadratic objective function (2.1) is convex.

Smallest enclosing ball as a convex program. In order to show that the smallest enclosing ball of a point set can be extracted from the solution of a suitable convex quadratic program, we use the Karush–Kuhn–Tucker conditions and the following geometric fact, which is interesting by itself.

8.7.3 Lemma. *Let $S = \{\mathbf{s}_1,\ldots,\mathbf{s}_k\} \subseteq \mathbb{R}^d$ be a set of points on the boundary of a ball B with center $\mathbf{s}^* \in \mathbb{R}^d$. Then the following two statements are equivalent.*

(i) *B is the unique smallest enclosing ball of S.*
(ii) *For every $\mathbf{u} \in \mathbb{R}^d$, there is an index $j \in \{1,2,\ldots,k\}$ such that*

$$\mathbf{u}^T(\mathbf{s}_j - \mathbf{s}^*) \leq 0.$$

It is a simple exercise to show that (ii) can be reexpressed as follows: There is no hyperplane that strictly separates S from $\mathbf{s}^*$. From the Farkas lemma (Section 6.4), one can in turn derive the following equivalent formulation: The point $\mathbf{s}^*$ is in the convex hull of the

points S. We thus have a simple geometric condition that character-
izes smallest enclosing balls in terms of their boundary points. From
the geometric intuition in the plane, this is quite plausible: If $\mathbf{s}^*$ is in
the convex hull of S, then $\mathbf{s}^*$ cannot be moved without making the
distance to at least one point larger. But if $\mathbf{s}^*$ is separated from S
by a hyperplane, then moving $\mathbf{s}^*$ toward this hyperplane results in an
enclosing ball of smaller radius. The direction $\mathbf{u}$ of movement satisfies
$\mathbf{u}^T(\mathbf{s}_j - \mathbf{s}^*) > 0$ for all j.

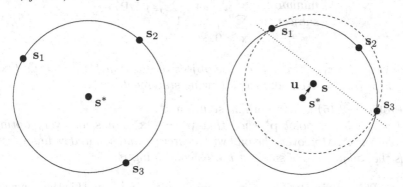

Proof. We start by analyzing the distance between a point $\mathbf{s}_j \in S$ and a
potential ball center $\mathbf{s} \neq \mathbf{s}^*$. Let r be the radius of the ball B. Given $\mathbf{s} \neq \mathbf{s}^*$,
we can uniquely write it in the form

$$\mathbf{s} = \mathbf{s}^* + t\mathbf{u},$$

where $\mathbf{u}$ is a vector of length 1 and $t > 0$. For $j = 1, 2, \ldots, k$ we get

$$
\begin{aligned}
(\mathbf{s}_j - \mathbf{s})^T (\mathbf{s}_j - \mathbf{s}) &= (\mathbf{s}_j - \mathbf{s}^* - t\mathbf{u})^T (\mathbf{s}_j - \mathbf{s}^* - t\mathbf{u}) \\
&= (\mathbf{s}_j - \mathbf{s}^*)^T (\mathbf{s}_j - \mathbf{s}^*) + t^2 \mathbf{u}^T \mathbf{u} - 2t\mathbf{u}^T (\mathbf{s}_j - \mathbf{s}^*) \\
&= r^2 + t^2 - 2t\mathbf{u}^T (\mathbf{s}_j - \mathbf{s}^*).
\end{aligned}
$$

Given $\alpha \in \mathbb{R}$ and sufficiently small $t > 0$, we have $t^2 - 2t\alpha > 0$ if and only if
$\alpha \leq 0$. This shows that (for sufficiently small t)

$$(\mathbf{s}_j - \mathbf{s})^T (\mathbf{s}_j - \mathbf{s}) > r^2 \quad \Leftrightarrow \quad \mathbf{u}^T (\mathbf{s}_j - \mathbf{s}^*) \leq 0, \tag{8.14}$$

where the implication "$\Leftarrow$" holds for all $t > 0$.

This equivalence implies the two directions of the lemma. For (i)$\Rightarrow$(ii), we
argue as follows: Since $\mathbf{s}^*$ is the unique point with distance at most r from
all points in S, we know that for every $\mathbf{u}$ with $\|\mathbf{u}\| = 1$ and for *all* $t > 0$,
the point $\mathbf{s} = \mathbf{s}^* + t\mathbf{u}$ has distance more than r to one of the points in S.
By the implication "$\Rightarrow$" of (8.14), there is some j with $\mathbf{u}^T (\mathbf{s}_j - \mathbf{s}^*) \leq 0$. To
show (ii)$\Rightarrow$(i), let us consider any point $\mathbf{s}$ of the form $\mathbf{s}^* + t\mathbf{u} \neq \mathbf{s}^*$. Since
there is an index j with $\mathbf{u}^T (\mathbf{s}_j - \mathbf{s}^*) \leq 0$, implication "$\Leftarrow$" of (8.14) shows

that $\mathbf{s}$ has distance more than r to some point in S. It follows that B is the unique smallest enclosing ball of S.   □

Now we can state and prove the main result.

8.7.4 Theorem. *Let* $\mathbf{p}_1, \ldots, \mathbf{p}_n$ *be points in* $\mathbb{R}^d$, *and let* Q *be the* $d \times n$ *matrix whose* j*th column is formed by the* d *coordinates of the point* $\mathbf{p}_j$. *Let us consider the optimization problem*

$$
\begin{array}{ll}
\text{minimize} & \mathbf{x}^T Q^T Q \mathbf{x} - \sum_{j=1}^n x_j \mathbf{p}_j^T \mathbf{p}_j \\
\text{subject to} & \sum_{j=1}^n x_j = 1 \\
& \mathbf{x} \geq \mathbf{0}
\end{array}
\tag{8.15}
$$

in the variables $x_1, \ldots, x_n$. *Then the objective function* $f(\mathbf{x}) := \mathbf{x}^T Q^T Q \mathbf{x} - \sum_{j=1}^n x_j \mathbf{p}_j^T \mathbf{p}_j$ *is convex, and the following statements hold.*

(i) *Problem (8.15) has an optimal solution* $\mathbf{x}^*$.

(ii) *There exists a point* $\mathbf{p}^*$ *such that* $\mathbf{p}^* = Q \mathbf{x}^*$ *holds for every optimal solution* $\mathbf{x}^*$. *Moreover, the ball with center* $\mathbf{p}^*$ *and squared radius* $-f(\mathbf{x}^*)$ *is the unique ball of smallest radius containing* P.

Proof. The matrix $Q^T Q$ is positive semidefinite, and from this the convexity of f is easy to verify (we leave it as an exercise).

The feasible region of program (8.15) is a compact set (actually, a *simplex*), and we are minimizing a continuous function over it. Consequently, there exists an optimal solution $\mathbf{x}^*$. In order to apply the Karush–Kuhn–Tucker conditions, we need the gradient of the objective function:

$$
\nabla f(\mathbf{x}) = 2\mathbf{x}^T Q^T Q - (\mathbf{p}_1^T \mathbf{p}_1, \mathbf{p}_2^T \mathbf{p}_2, \ldots, \mathbf{p}_n^T \mathbf{p}_n).
$$

The program has only one equality constraint. With $\mathbf{p}^* = Q\mathbf{x}^* = \sum_{j=1}^n x_j^* \mathbf{p}_j$, Proposition 8.7.2 tells us that we find a 1-dimensional vector $\tilde{\mathbf{y}} = (\mu)$ such that

$$
2\mathbf{p}_j^T \mathbf{p}^* - \mathbf{p}_j^T \mathbf{p}_j + \mu \left\{ \begin{array}{ll} = 0 & \text{if } x_j^* > 0 \\ \geq 0 & \text{otherwise.} \end{array} \right.
\tag{8.16}
$$

Subtracting $\mathbf{p}^{*T} \mathbf{p}^*$ from both sides and multiplying by -1 yields

$$
\|\mathbf{p}_j - \mathbf{p}^*\|^2 \left\{ \begin{array}{ll} = \mu + \mathbf{p}^{*T} \mathbf{p}^* & \text{if } x_j^* > 0 \\ \leq \mu + \mathbf{p}^{*T} \mathbf{p}^* & \text{otherwise.} \end{array} \right.
$$

This means that $\mathbf{p}^*$ is the center of a ball of radius $r = \sqrt{\mu + \mathbf{p}^{*T} \mathbf{p}^*}$ that contains all points from P and has the points $\mathbf{p}_j$ with $x_j^* > 0$ on the boundary. From (8.16) and $\mathbf{x} \geq \mathbf{0}$ we also get

$$
\mu = \sum_{j=1}^n x_j^* \mu = \sum_{j=1}^n x_j^* \mathbf{p}_j^T \mathbf{p}_j - 2 \sum_{j=1}^n x_j^* \mathbf{p}_j^T \mathbf{p}^* = \sum_{j=1}^n x_j^* \mathbf{p}_j^T \mathbf{p}_j - 2\mathbf{p}^{*T} \mathbf{p}^*,
$$

and $r^2 = \sum_{j=1}^{n} x_j^* \mathbf{p}_j^T \mathbf{p}_j - \mathbf{p}^{*T} \mathbf{p}^* = -f(\mathbf{x}^*)$ follows.

It remains to prove that there can be no other ball of radius at most r that contains all points from P (this also shows that $\mathbf{p}^*$ does not depend on the choice of $\mathbf{x}^*$).

We define $F = \{j \in \{1, 2, \ldots, n\} : x_j^* > 0\}$ and apply Lemma 8.7.3 with $\mathbf{s}^* = \mathbf{p}^*$ and

$$S = \{\mathbf{p}_j : j \in F\}.$$

We already know that these points are on the boundary of a ball B of radius r around $\mathbf{p}^* = \sum_{j \in F} x_j^* \mathbf{p}_j$. Using $\sum_{j \in F} x_j^* = 1$, we get that the following holds for all vectors $\mathbf{u}$:

$$\sum_{j \in F} x_j^* \mathbf{u}^T (\mathbf{p}_j - \mathbf{p}^*) = \mathbf{u}^T \left(\sum_{j \in F} x_j^* \mathbf{p}_j - \sum_{j \in F} x_j^* \mathbf{p}^* \right) = \mathbf{u}^T (\mathbf{p}^* - \mathbf{p}^*) = 0.$$

It follows that there must be some $j \in F$ with $\mathbf{u}^T (\mathbf{p}_j - \mathbf{p}^*) \leq 0$. By Lemma 8.7.3, B is the unique smallest enclosing ball of $S \subseteq P$, and this implies that B is the unique smallest enclosing ball of P as well. $\qquad\square$

A recent book on the topics of this section is

S. Boyd and L. Vandenberghe: *Convex Optimization*, Cambridge University Press, Cambridge 2004.

9. Software and Further Reading

LP-solvers. The most famous (and expensive) software package for solving linear programs and integer programs is called CPLEX. Freely available codes with similar functionality, although not quite as strong as CPLEX, are lp_solve, GLPK, and CLP. The website

www-neos.mcs.anl.gov/neos

contains a guide to many other optimization software systems, and it also provides an overview of web solvers, to which one can send an input of an optimization problem and, with a bit of luck, be returned an optimum.

The *computational geometry algorithms library* CGAL (www.cgal.org) contains software for solving linear and convex quadratic programs using exact rational arithmetic. We refer to the website of this book (http://www.inf.ethz.ch/personal/gaertner/lpbook) for further information.

Books. The web bookstore Amazon offers more books with "linear programming" in the title than with "astrology," and so it is clear that we can mention only a very narrow selection from the literature.

A reasonably recent, accessible, and quite comprehensive (but not exactly cheap) textbook of linear programming is

> D. Bertsimas and J. Tsitsiklis: *Introduction to Linear Optimization,*
> Athena Scientific, Belmont, Massachusetts, 1997.

Both linear and integer programming are treated on an excellent theoretical level in

> A. Schrijver: *Theory of Linear and Integer Programming,* Wiley-Interscience, New York 1986.

The book

> V. Chvátal: *Linear Programming,* W. H. Freeman, New York 1983,

was considered one of the best textbooks in its time and it is still used widely. And those liking classical sources may appreciate

> G. B. Dantzig: *Linear Programming and Extensions,* Princeton University Press, Princeton 1963.

Appendix: Linear Algebra

Here we summarize facts and concepts from linear algebra used in this book. This part is not meant to be a textbook introduction to the subject. It is mainly intended for the reader who has some knowledge of the area but may have forgotten the exact definitions or may know them in a slightly different form. The number of introductory textbooks is vast; in order to cite at least one, we mention

> G. Strang: *Introduction to Linear Algebra*, 3rd edition, Wellesley-Cambridge Press, Wellesley, Massachusetts, 2003.

Vectors. We work exclusively with vectors in $\mathbb{R}^n$, and so for us, a **vector** is an ordered n-tuple $\mathbf{v} = (v_1, v_2, \ldots, v_n) \in \mathbb{R}^n$ of real numbers. We denote vectors by boldface letters, and v_i denotes the ith component of $\mathbf{v}$. For $\mathbf{u}, \mathbf{v} \in \mathbb{R}^n$ we define the sum componentwise:

$$\mathbf{u} + \mathbf{v} = (u_1 + v_1, u_2 + v_2, \ldots, u_n + v_n).$$

The multiplication of $\mathbf{v} \in \mathbb{R}^n$ by a real number t is also given componentwise, by

$$t\mathbf{v} = (tv_1, tv_2, \ldots, tv_n).$$

We use the notation $\mathbf{0}$ for the zero vector, with all components 0, and $\mathbf{1}$ denotes a vector with all components equal to 1.

A **linear subspace** (or a *vector subspace*) of $\mathbb{R}^n$ is a set $V \subseteq \mathbb{R}^n$ that contains $\mathbf{0}$ and is closed under addition and multiplication by a real number; that is, if $\mathbf{u}, \mathbf{v} \in V$ and $t \in \mathbb{R}$, we have $\mathbf{u} + \mathbf{v} \in V$ and $t\mathbf{v} \in V$. For example, the linear subspaces of $\mathbb{R}^3$ are $\{\mathbf{0}\}$, lines passing through $\mathbf{0}$, planes passing through $\mathbf{0}$, and $\mathbb{R}^3$ itself. An **affine subspace** is any set of the form $\mathbf{u} + V = \{\mathbf{u} + \mathbf{v} : \mathbf{v} \in V\} \subseteq \mathbb{R}^n$, where V is a linear subspace of $\mathbb{R}^n$ and $\mathbf{u} \in \mathbb{R}^n$. Geometrically, it is a linear subspace translated by some fixed vector. The affine subspaces of $\mathbb{R}^3$ are all the one-point subsets, all the lines, all the planes, and $\mathbb{R}^3$.

A **linear combination** of vectors $\mathbf{v}_1, \mathbf{v}_2, \ldots, \mathbf{v}_m \in \mathbb{R}^n$ is any vector of the form $t_1\mathbf{v}_1 + t_2\mathbf{v}_2 + \cdots + t_m\mathbf{v}_m$, where $t_1, \ldots, t_m$ are real numbers. Vectors $\mathbf{v}_1, \mathbf{v}_2, \ldots, \mathbf{v}_m \in \mathbb{R}^n$ are **linearly independent** if the only linear combination of $\mathbf{v}_1, \mathbf{v}_2, \ldots, \mathbf{v}_m$ equal to $\mathbf{0}$ has $t_1 = t_2 = \cdots = t_m = 0$. Equivalently, linear independence means that none of the $\mathbf{v}_i$ can be expressed as a linear combination of the remaining ones.

The **linear span** of a set $X \subseteq \mathbb{R}^n$ is the smallest (with respect to inclusion) linear subspace of $\mathbb{R}^n$ that contains X. Explicitly, it is the set of all linear combinations of finitely many vectors from X. The linear span of any set, even the empty one, always contains $\mathbf{0}$, which is formally considered as a linear combination of the empty set of vectors.

A **basis** of a linear subspace $V \subseteq \mathbb{R}^n$ is a linearly independent set of vectors from V whose linear span is V. The **standard basis** of $\mathbb{R}^n$ consists of the vectors $\mathbf{e}_1, \mathbf{e}_2, \ldots, \mathbf{e}_n$, where $\mathbf{e}_i$ is the vector with 1 at the ith position and 0's elsewhere.

All bases of a given linear subspace V have the same number of vectors, and this number $\dim(V)$ is the **dimension** of V. In particular, each basis of $\mathbb{R}^n$ has n vectors.

Matrices. A matrix is a rectangular table of numbers (real numbers, in our case). An $m \times n$ matrix has m rows and n columns. If a matrix is called A, then its entry in the ith row and jth column is usually denoted by a_{ij}. So, for example, a 3×4 matrix A has the general form

$$\begin{pmatrix} a_{11} & a_{12} & a_{13} & a_{14} \\ a_{21} & a_{22} & a_{23} & a_{24} \\ a_{31} & a_{32} & a_{33} & a_{34} \end{pmatrix}.$$

A matrix is denoted by writing large parentheses to enclose the table of elements. A **submatrix** of a matrix A is any matrix that can be obtained from A by deleting some rows and some columns (including A itself, where we delete nothing).

A matrix is multiplied by a number t by multiplying each entry by t. Two $m \times n$ matrices A and B are added by adding the corresponding entries. That is, if we set $C = A + B$, we have $c_{ij} = a_{ij} + b_{ij}$ for $i = 1, 2, \ldots, m$ and $j = 1, 2, \ldots, n$.

Matrix multiplication is more complicated. A product AB, where A and B are matrices, is defined only if the number of columns of A is the same as the number of rows of B. If A is an $m \times n$ matrix and B is an $n \times p$ matrix, then the product $C = AB$ is an $m \times p$ matrix given by

$$c_{ij} = a_{i1}b_{1j} + a_{i2}b_{2j} + \cdots + a_{in}b_{nj}.$$

Pictorially,

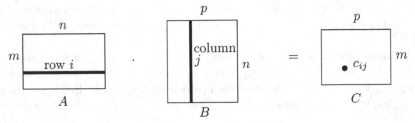

Matrix multiplication is *associative*, meaning that $A(BC) = (AB)C$ whenever at least one of the two sides is defined, and this is quite often used in proofs. We also recall the well-known fact that in general, matrix multiplication is not commutative; i.e., typically $AB \neq BA$.

We also multiply matrices and vectors. In such context, a vector $\mathbf{x} \in \mathbb{R}^n$ is usually considered as an $n \times 1$ matrix; thus, in the matrix form, a vector $\mathbf{x} = (x_1, x_2, \ldots, x_n)$ should be written as a column:

$$\begin{pmatrix} x_1 \\ x_2 \\ \vdots \\ x_n \end{pmatrix}.$$

So if A is an $m \times n$ matrix and $\mathbf{x} \in \mathbb{R}^n$ is a vector, then the product $A\mathbf{x}$ is a vector in $\mathbb{R}^m$.

This is used in the matrix notation $A\mathbf{x} = \mathbf{b}$ for a system of linear equations. Here A is a given $m \times n$ matrix, $\mathbf{b} \in \mathbb{R}^m$ is a given vector, and $\mathbf{x} = (x_1, x_2, \ldots, x_n)$ is a vector of n unknowns. So $A\mathbf{x} = \mathbf{b}$ is a shorthand for the system of m equations

$$
\begin{array}{ccccccc}
a_{11}x_1 & + & a_{12}x_2 & + \cdots + & a_{1n}x_n & = & b_1 \\
a_{21}x_1 & + & a_{22}x_2 & + \cdots + & a_{2n}x_n & = & b_2 \\
& & \vdots & & \vdots & & \vdots \\
a_{m1}x_1 & + & a_{m2}x_2 & + \cdots + & a_{mn}x_n & = & b_m.
\end{array}
$$

If A is an $m \times n$ matrix, then A^T denotes the $n \times m$ matrix having the element a_{ji} in the ith row and jth column. The matrix A^T is called the **transpose** of the matrix A. For transposing the matrix product, we have the formula $(AB)^T = B^T A^T$.

A **square matrix** is an $n \times n$ matrix, i.e., one with the same number of rows and columns. A **diagonal matrix** is a square matrix D with $d_{ij} = 0$ for all $i \neq j$; that is, it may have nonzero elements only on the diagonal. The $n \times n$ **identity matrix** I_n has 1's on the diagonal and 0's elsewhere. For any $m \times n$ matrix A we have $I_m A = A I_n = A$.

Rank, inverse, and linear systems. Each row of an $m \times n$ matrix A can also be regarded as a vector in $\mathbb{R}^n$. The linear span of all rows of A is a subspace of $\mathbb{R}^n$ called the **row space** of A, and similarly, we define the **column space** of A as the linear subspace of $\mathbb{R}^m$ spanned by the columns of A. An important result of linear algebra tells us that for every matrix A the row space and the column space have the same dimension, and this dimension is called the **rank** of A. In particular, an $m \times n$ matrix A has rank m if and only if the rows of A are linearly independent (which can happen only if $m \leq n$). An $n \times n$ matrix is called **nonsingular** if it has rank n; otherwise, it is singular.

Let A be a square matrix. A matrix B is called an **inverse** of A if $AB = I_n$. An inverse to A exists if and only if A is nonsingular. In this case it is determined uniquely, it is denoted by A^{-1}, and it is inverse from both sides; that is, $AA^{-1} = A^{-1}A = I_n$. For the inverse of a product we have $(AB)^{-1} = B^{-1}A^{-1}$.

Let us again consider a system $A\mathbf{x} = \mathbf{b}$ of m linear equations with n unknowns. For $\mathbf{b} = \mathbf{0}$, the set of all solutions is easily seen to be a linear subspace of $\mathbb{R}^n$. Its dimension equals n minus the rank of A. In particular, for $m = n$ (i.e., A is a square matrix), the system $A\mathbf{x} = \mathbf{0}$ has $\mathbf{x} = \mathbf{0}$ as the only solution if and only if A is nonsingular.

For $\mathbf{b} \neq \mathbf{0}$, the system $A\mathbf{x} = \mathbf{b}$ may or may not have a solution. If it has at least one solution, then the solution set is an affine subspace of $\mathbb{R}^n$, again of dimension n minus the rank of A. If $m = n$ and A is nonsingular, then $A\mathbf{x} = \mathbf{b}$ always has exactly one solution.

Row operations and Gaussian elimination. By an **elementary row operation** we mean one of the following two operations on a given matrix A:

(a) Multiplying all entries in some row of A by a *nonzero* real number t.
(b) Replacing the ith row of A by the sum of the ith row and jth row for some $i \neq j$.

Gaussian elimination is a systematic procedure that, given an input matrix A, performs a sequence of elementary row operations on it so that it is converted to a *row echelon form*. This means that the resulting matrix looks as in the following picture:

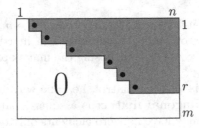

(the dots denote nonzero elements and the white region contains only 0's). In words, there exists an integer r such that the rows 1 through r are nonzero, the remaining rows are all zero, and if $j(i) = \min\{j : a_{ij} \neq 0\}$, then $j(1) < j(2) < \cdots < j(r)$.

The rank of a matrix in this form is clearly r, and since elementary row operations preserve rank, this procedure can be used to compute the rank. It is also commonly used for finding all solutions to the system $A\mathbf{x} = \mathbf{b}$: In this case, the matrix $(A\,|\,\mathbf{b})$ (the matrix A with $\mathbf{b}$ appended as the last column) is converted to row echelon form, and from this, the solution set can be computed easily. A variant of Gaussian elimination can also be used to compute the inverse matrix, essentially by solving the n linear systems $A\mathbf{x} = \mathbf{e}_i$, $i = 1, 2, \ldots, n$.

Determinants. Every square matrix A is assigned a number $\det(A)$ called the **determinant** of A. The determinant of A is defined by the formula

$$\det(A) = \sum_{\pi \in S_n} \operatorname{sgn}(\pi) \prod_{i=1}^{n} a_{i,\pi(i)},$$

where the sum is over all permutations π of the set $\{1, 2, \ldots, n\}$ and where $\operatorname{sgn}(\pi)$ denotes the *sign* of a permutation π. The sign of any permutation is either $+1$ or -1, and it can be compactly defined as the sign of the expression

$$\prod_{1 \le i < j \le n} \big(\pi(j) - \pi(i)\big).$$

For example, the determinant of a 2×2 matrix A equals $a_{11}a_{22} - a_{12}a_{21}$.

We have $\det(A) \ne 0$ if and only if A is nonsingular. **Cramer's rule** is a formula describing the (unique) solution of the linear system $A\mathbf{x} = \mathbf{b}$ with A square and nonsingular. It asserts that

$$x_j = \frac{\det(A_{j \to \mathbf{b}})}{\det(A)},$$

where $A_{j \to \mathbf{b}}$ denotes the matrix A with the jth column replaced by the vector $\mathbf{b}$.

For any column index j, we have the following formula (the **Laplace expansion** of the determinant according to a column):

$$\det(A) = \sum_{i=1}^{n} (-1)^{i+j} a_{ij} \det(A^{ij}),$$

where A^{ij} denotes the matrix arising from A by deleting the ith row and the jth column.

Scalar product, Euclidean norm, orthogonality. The (standard) **scalar product** of two vectors $\mathbf{x}, \mathbf{y} \in \mathbb{R}^n$ is the number $x_1 y_1 + x_2 y_2 + \cdots + x_n y_n$. We often write the scalar product as $\mathbf{x}^T \mathbf{y}$, although formally, $\mathbf{x}^T \mathbf{y}$ is a 1×1 matrix whose single entry is the scalar product. The **Euclidean norm** of a vector $\mathbf{x} \in \mathbb{R}^n$ is denoted by $\|\mathbf{x}\|$ and defined by $\|\mathbf{x}\| = \sqrt{\mathbf{x}^T \mathbf{x}} = \sqrt{x_1^2 + \cdots + x_n^2}$. The **Euclidean distance** of two vectors $\mathbf{x}$ and $\mathbf{y}$ is $\|\mathbf{x} - \mathbf{y}\|$.

Two vectors $\mathbf{x}, \mathbf{y} \in \mathbb{R}^n$ are called **orthogonal** if $\mathbf{x}^T \mathbf{y} = 0$. More generally, the **angle** of nonzero vectors $\mathbf{x}, \mathbf{y} \in \mathbb{R}^n$ is defined as the angle between 0 and π whose cosine equals $\frac{\mathbf{x}^T \mathbf{y}}{\|\mathbf{x}\| \cdot \|\mathbf{y}\|}$.

A square matrix A is called **orthogonal** if each column has Euclidean norm 1 and every two distinct columns are orthogonal. Equivalently, we have $A^{-1} = A^T$. From this one can also see that A is orthogonal if and only if A^T is.

The **orthogonal complement** of a set $X \subseteq \mathbb{R}^n$ is the set $X^\perp = \{\mathbf{y} \in \mathbb{R}^n : \mathbf{x}^T\mathbf{y} = 0$ for all $\mathbf{x} \in X\}$. It is always a linear subspace of $\mathbb{R}^n$. If V is a linear subspace of $\mathbb{R}^n$, then $\dim(V) + \dim(V^\perp) = n$.

If V is a linear subspace of $\mathbb{R}^n$, the **orthogonal projection** on V is a mapping $\mathbb{R}^n \to V$ that assigns to each $\mathbf{x} \in \mathbb{R}^n$ a vector $\mathbf{y} \in V$ such that $\mathbf{x} - \mathbf{y}$ is orthogonal to all vectors in V. It can be shown that for every $\mathbf{x}$ such a $\mathbf{y}$ is unique and it is also the point of V that minimizes the Euclidean distance $\mathbf{x} - \mathbf{y}$ among all points of V.

Glossary or: What Was Neglected

The theory of linear programming is huge, and many interesting things were not addressed in this book. In the glossary we briefly explain some of the common terms found elsewhere. This should help the reader in reading more advanced sources, say research papers, or in a conversation at a linear programming banquet. Our coverage is by no means complete, and to some extent, it is also guided by personal taste.

Bounds for the variables. Linear programs occurring in practice often prescribe upper and lower bounds for (some of) the variables. For simplicity, let us assume that we are dealing with a linear program of the form

$$\begin{aligned} \text{maximize} \quad & \mathbf{c}^T\mathbf{x} \\ \text{subject to} \quad & A\mathbf{x} = \mathbf{b} \\ & \mathbf{0} \leq \mathbf{x} \leq \mathbf{u}, \end{aligned}$$

where $u_j \in \mathbb{R} \cup \{\infty\}$ for $j = 1, 2, \ldots, n$.

We have seen in Section 4.1 how to convert this linear program into equational form, and after this conversion, we can solve it using the simplex method. The problem here is that the conversion generates many new variables and constraints, making the problem substantially larger and the computations within a single pivot step more expensive.

A better way is to handle the constraints $\mathbf{x} \leq \mathbf{u}$ implicitly in the simplex method. This is easy: During the selection of the leaving variable we also have to consider the possibility that the entering variable reaches its upper bound *before* any basic variable reaches one of its bounds. In this situation, the entering variable is simply set to its upper bound, and the basis B does not change at all. In general, every nonbasic variable attains one of its bounds, and "entering it" means to let its value change in the direction of the other bound. As before, if $x^*_{\ell_j} = 0$ in the current basic feasible solution, then x_{ℓ_j} is a candidate for the entering variable if and only if the corresponding coefficient r_j is positive. On the other hand, if $x^*_{\ell_j} = u_j$, then x_{ℓ_j} is a candidate if and only if $r_j < 0$.

This scheme easily generalizes to bounds of the form $\boldsymbol{\ell} \leq \mathbf{x} \leq \mathbf{u}$.

Branch and bound is a general method for solving optimization problems by a clever enumeration of possible solutions. Here we describe how it works for integer programs (Chapter 3). We suppose that the integer program is

$$\text{maximize } \mathbf{c}^T \mathbf{x} \text{ subject to } A\mathbf{x} \le \mathbf{b}, \ \mathbf{x} \in \mathbb{Z}^n, \qquad \text{(ILP)}$$

and, for simplicity, that the polyhedron $P = \{\mathbf{x} \in \mathbb{R}^n : A\mathbf{x} \le \mathbf{b}\}$ is bounded. As a first step, we solve the LP relaxation obtained by dropping the integrality constraints $\mathbf{x} \in \mathbb{Z}^n$. If the LP relaxation is infeasible, then (ILP) is infeasible as well and we stop. Otherwise, the LP relaxation has an optimal solution $\mathbf{x}^*$. If $\mathbf{x}^* \in \mathbb{Z}^n$, we have already solved the program (ILP), and if $\mathbf{x}^* \notin \mathbb{Z}^n$, we chose some nonintegral component x_j^* and "split" the integer program into two integer programs: The first one, (ILP$_\le$), is obtained from (ILP) by adding the constraint $x_j \le \lfloor x_j^* \rfloor$, while the second one, (ILP$_\ge$), arises from (ILP) by adding the constraint $x_j \ge \lfloor x_j^* \rfloor + 1$. This is the *branching step*.

Since every feasible solution of (ILP) is feasible for exactly one of the programs (ILP$_\le$) and (ILP$_\ge$), we have only to solve the latter two integer programs to find an optimal solution of (ILP), if it exists. The "only" refers to the fact that both of these programs have strictly smaller feasible regions than the original one. To solve (ILP$_\le$), say, we proceed in the same fashion (solve the LP relaxation, split into two subproblems if necessary).

In this way, we explore an implicitly given binary tree whose nodes correspond to subproblems of the original integer program. The exploration stops at a node when the LP relaxation becomes infeasible or has an integral optimal solution. From the assumption that P is bounded it is not hard to show that this eventually happens along every exploration path. Therefore, the process terminates and computes an optimal solution of the "root program" (ILP), if there is one.

So far this approach is missing the advertised cleverness, but here it comes: Whenever an integral solution of a subproblem has been discovered, its value (of the objective function) is a lower bound for the optimal value of (ILP). During exploration, we maintain the highest such lower bound z^*. If at some subsequent node the optimal solution of the LP relaxation has objective function value at most z^*, we can conclude that the subtree below that node need not be explored: No integral solution obtained from it can beat our current best solution with value z^*. This is the bounding step.

In the worst case the bounding step may not prune any subtree, but in many practical applications it results in enormous savings and allows for solving large integer programs. The effectiveness of branch and bound also depends on the choice of the nonintegral components x_j^* at the nodes, and on the order in which nodes are explored. More generally, the "axis-parallel" split according to the constraints $x_j \le \lfloor x_j^* \rfloor$ and $x_j \ge \lfloor x_j^* \rfloor + 1$ may be replaced by a split along an arbitrary direction (branching on hyperplanes).

Branch and cut. This method combines the branch and bound technique with cutting planes. In many cases, branch and bound does not work well since the plain LP relaxation does not provide a useful upper bound for the optimal solution of an integer program (we have seen a simple example for that in Section 3.4). Before branching, one therefore tries to get a better upper bound by adding cutting planes to the LP relaxation. Sometimes, these are tailor-made for the problem at hand and cut off many more fractional solutions than general-purpose cutting planes like Gomory cuts. The branching starts only after all cutting planes from a predetermined set of candidate inequalities have been added. In the subproblems one proceeds similarly.

Chvátal rank. We consider a polyhedron $P \subseteq \mathbb{R}^n$. If $\mathbf{a}^T\mathbf{x} \le b$, with $\mathbf{a} \in \mathbb{Z}^n$ and $b \in \mathbb{Q}$, is some inequality satisfied by all $\mathbf{x} \in P$, then the inequality

$$\mathbf{a}^T\mathbf{x} \le \lfloor b \rfloor$$

is satisfied by all integral points in P. This inequality is a *Chvátal–Gomory cut* of P.

Let us define P' as the set of points that satisfy all Chvátal–Gomory cuts of P. It follows that $P' \supseteq P_I$, where P_I is the convex hull of all integral points in P, the *integer hull* of P.

Let us now assume that P is a *rational polyhedron*, i.e., one that can be described by an inequality system $A\mathbf{x} \le \mathbf{b}$ with all components of A and $\mathbf{b}$ rational. In this case, one can show that P' is again a polyhedron. Moreover, it is easy to see that $P' \subseteq P$. We thus have

$$P = P^{(0)} \supseteq P^{(1)} \supseteq P^{(2)} \supseteq \cdots \supseteq P_I,$$

where $P^{(k)} = (P^{(k-1)})'$ for $k > 0$.

It is known that there is a finite number t such that $P^{(t)} = P_I$; the smallest such t is the Chvátal rank of the rational polyhedron P. It is a measure of "nonintegrality" of P. Such a number t even exists if P is nonrational but bounded.

Column generation. We have pointed out in Section 7.1 that even linear programs with a very large (possibly infinite) number of constraints can efficiently be solved, by the ellipsoid method or interior point methods, if a separation oracle is available. Given a candidate solution $\mathbf{s}$, such an oracle either certifies that $\mathbf{s}$ is feasible, or it returns a violated constraint. Let us consider the dual scenario—a linear program with a very large number of variables. Even if this linear program is too large to be explicitly stored, we may be able to solve it using the simplex method. The crucial observation is that in every pivot step, the simplex method needs just *one* entering variable (and the tableau column associated with it) to proceed; see Section 5.6. If we have an *improvement oracle* that returns such a variable (or certifies that the current basic feasible solution is optimal), we can still use the simplex method to solve the linear program. This method is called *(delayed) column generation*. It can be used in applications as in Section 2.7 (paper cutting), for example, where the number of variables can quickly become very large. In fact, since an improvement oracle can be interpreted as a separation oracle for the dual linear program, we can also use the ellipsoid method to solve the linear program, using only a polynomial number of calls to the improvement oracle.

Complementary slackness.   The following corollary of the duality theorem is known as the theorem of *complementary slackness*.

Let $\mathbf{x}^* = (x_1^*, \ldots, x_n^*)$ be a feasible solution of the linear program

$$\text{maximize } \mathbf{c}^T \mathbf{x} \text{ subject to } A\mathbf{x} \leq \mathbf{b} \text{ and } \mathbf{x} \geq \mathbf{0}, \qquad (P)$$

and let $\mathbf{y}^* = (y_1^*, \ldots, y_m^*)$ be a feasible solution of the dual linear program

$$\text{minimize } \mathbf{b}^T \mathbf{y} \text{ subject to } A^T \mathbf{y} \geq \mathbf{c} \text{ and } \mathbf{y} \geq \mathbf{0}. \qquad (D)$$

Then the following two statements are equivalent:
(i) $\mathbf{x}^*$ is optimal for (P) and $\mathbf{y}^*$ is optimal for (D).
(ii) For all $i = 1, 2, \ldots, m$, $\mathbf{x}^*$ satisfies the ith constraint of (P) with equality or $y_i^* = 0$; similarly, for all $j = 1, 2, \ldots, n$, $\mathbf{y}^*$ satisfies the jth constraint of (D) with equality or $x_j^* = 0$.

In words, statement (ii) means the following: If we pair up each (primal or dual) nonnegativity constraint with its corresponding (dual or primal) inequality, then at least one of the constraints in each pair is satisfied with equality ("has no slack") at $\mathbf{x}^*$ or $\mathbf{y}^*$.

Complementary slackness is often encountered as a "combinatorial" proof of optimality, as opposed to the "numerical" proof obtained by comparing the values of the objective functions. We have come across complementary slackness at various places, without calling it so: in the physical interpretation of duality (Section 6.2), in connection with the primal–dual central path (Section 7.2), and in the Karush–Kuhn–Tucker conditions (Section 8.7).

Criss–cross method.   This is yet another method for solving linear programs in equational form. Like the simplex method and the dual simplex method, it goes through a sequence of simplex tableaus

$$\begin{array}{rcl} \mathbf{x}_B & = & \mathbf{p} \;+\; Q\,\mathbf{x}_N \\ \hline z & = & z_0 \;+\; \mathbf{r}^T \mathbf{x}_N \end{array}$$

The criss–cross method only requires the set B to be a basis, meaning that the submatrix A_B is regular. This property allows us to form the tableau, but it guarantees neither $\mathbf{p} \geq \mathbf{0}$ (as in the simplex method) nor $\mathbf{r} \leq \mathbf{0}$ (as in the dual simplex method).

The criss–cross method has two types of pivot steps. A *primal* pivot step starts by choosing $\beta \in \{1, 2, \ldots, n - m\}$ and $\alpha \in \{1, 2, \ldots, m\}$ such that $r_\beta > 0$ and $q_{\alpha\beta} < 0$. This is as in the simplex method, except that α does not have to satisfy a minimum ratio condition as in equation (5.3) of Section 5.6. The absence of this condition is explained by the fact that the criss–cross method does not have to maintain feasibility ($\mathbf{p} \geq \mathbf{0}$).

In a *dual* pivot step, $\alpha \in \{1, 2, \ldots, m\}$ and $\beta \in \{1, 2, \ldots, n - m\}$ are chosen such that $p_\alpha < 0$ and $q_{\alpha\beta} > 0$. This is as in the dual simplex method, again without any minimum-ratio condition, since the criss–cross method does not maintain dual feasibility ($\mathbf{r} \leq \mathbf{0}$) either.

In both situations, $B' = (B \setminus \{k_\alpha\}) \cup \{\ell_\beta\}$ is the next basis, where $q_{\alpha\beta} \neq 0$ guarantees that B' is indeed a basis (see the proof of Lemma 5.6.1).

If the linear program has an optimal solution, it can be shown that a basis either induces an optimal basic feasible solution ($\mathbf{p} \geq \mathbf{0}, \mathbf{r} \leq \mathbf{0}$) or allows for a primal or a dual pivot step. This means that the criss–cross method cannot get stuck. It can cycle, though, even if the linear program is not degenerate.

However, there is a pivot rule for the criss–cross method that does not cycle. This rule is reminiscent of Bland's rule for the simplex method and goes as follows. We let α^* be the smallest value of α in any possible primal pivot step and β^* the smallest value of β in any possible dual pivot step. If $\alpha^* \leq \beta^*$, then we perform a primal pivot step with $\alpha = \alpha^*$ and β as small as possible, and otherwise, we perform a dual pivot step with $\beta = \beta^*$ and α as small as possible.

Despite its simplicity and the fact that the computations can start from any basis (there is no need for an auxiliary problem), the criss–cross method is not used in practice, since it is slow. The feature that makes it theoretically appealing is that the computations depend only on the signs of the involved numbers, but not on their magnitudes. This allows the method to be generalized to situations beyond linear programming, where no concept of magnitude exists, such as "linear programming" over oriented matroids.

Cutting plane. Given a system S of linear inequalities, another linear inequality is called a cutting plane for S if it is satisfied by all integral solutions of S, but it is violated by some non-integral solution of S. A cutting plane for S exists if and only if the polyhedron corresponding to the S is nonintegral. The cutting plane "cuts off" a part of this polyhedron that is free of integer points.

Dantzig–Wolfe decomposition. Often the constraint matrix A of a linear program has a special structure that can be exploited in order to solve the problem faster. The Dantzig–Wolfe decomposition is a particular technique to do this within the framework of the simplex method. Given a linear program

$$\text{maximize } \mathbf{c}^T \mathbf{x} \text{ subject to } A\mathbf{x} = \mathbf{b} \text{ and } \mathbf{x} \geq \mathbf{0}$$

in equational form, we partition the m equations into a system $A'\mathbf{x} = \mathbf{b}'$ of "difficult" equations and a system $A''\mathbf{x} = \mathbf{b}''$ of "easy" equations. How this is done depends on the concrete problem, but as a general guideline, the system $A'\mathbf{x} = \mathbf{b}'$ should contain the equations that involve many variables ("global constraints"), while the equations of $A''\mathbf{x} = \mathbf{b}''$ are the ones with few variables ("local constraints"). Often there are only few global constraints and the local constraints consist of independent blocks in the sense that constraints in different blocks do not share variables. Let us assume for simplicity that the polyhedron $P = \{\mathbf{x} \in \mathbb{R}^n : A''\mathbf{x} = \mathbf{b}'', \mathbf{x} \geq \mathbf{0}\}$ is bounded. Then we know that every $\mathbf{x} \in P$ is a convex combination of the vertices $\mathbf{v}_1, \mathbf{v}_2, \ldots, \mathbf{v}_K$ of P (Section 4.4). If we assume that we explicitly know these vertices, our linear program can be rewritten as a problem in K variables $t_1, t_2, \ldots, t_K$ as follows:

$$
\begin{aligned}
\text{Maximize} \quad & \sum_{j=1}^{K} t_j \mathbf{c}^T \mathbf{v}_j \\
\text{subject to} \quad & \sum_{j=1}^{K} t_j A' \mathbf{v}_j = \mathbf{b}' \\
& \sum_{j=1}^{K} t_j = 1 \\
& t_j \geq 0 \text{ for all } j = 1, 2, \ldots, K.
\end{aligned}
$$

This linear program typically has many fewer constraints than the one we started with, but many more variables. The smaller number of constraints is an advantage, since then the linear equation systems that have to be solved during the pivot steps of the simplex method are smaller as well. However, the resulting savings usually do not justify the large number of new variables. The trick that makes the approach efficient is to apply column generation: We can find an entering variable (and its associated tableau column) without precomputing the vertices $\mathbf{v}_j$. Indeed, a vertex of P that yields an entering variable can be obtained as a basic feasible optimal solution of a suitable linear program with the "easy" set of constraints $A''\mathbf{x} = \mathbf{b}''$ and $\mathbf{x} \geq \mathbf{0}$.

Devex is a pivot rule that efficiently approximates the STEEPEST EDGE pivot rule (Section 5.7), which in itself is somewhat costly to implement. Let us first recall STEEPEST EDGE: We want to choose the entering variable in such a way that the expression

$$
\frac{\mathbf{c}^T (\mathbf{x}_{\text{new}} - \mathbf{x}_{\text{old}})}{\|\mathbf{x}_{\text{new}} - \mathbf{x}_{\text{old}}\|}
$$

is maximized, where $\mathbf{x}_{\text{old}}$ and $\mathbf{x}_{\text{new}}$ are the current basic feasible solution and the next one, respectively. As usual, we let $\mathbf{p}, Q, \mathbf{r}, z_0$ be the parameters of the simplex tableau corresponding to the current feasible basis B, and we write $B = \{k_1, k_2, \ldots, k_m\}$ and $N = \{1, 2 \ldots, n\} \setminus B = \{\ell_1, \ell_2, \ldots, \ell_{n-m}\}$, where $k_1 < k_2 < \cdots < k_m$ and $\ell_1 < \ell_2 < \cdots < \ell_{n-m}$ (see Section 5.5).

Let us assume that the entering variable is x_v, and that $v = \ell_\beta$. Moreover, let us suppose that x_v has value $t \geq 0$ in $\mathbf{x}_{\text{new}}$.

With $\tilde{\mathbf{x}} = \mathbf{x}_{\text{old}}$, we then have

$$\mathbf{c}^T(\mathbf{x}_{\text{new}} - \mathbf{x}_{\text{old}}) = \mathbf{r}^T(\tilde{\mathbf{x}}_N + t\mathbf{e}_\beta - \tilde{\mathbf{x}}_N) = tr_\beta \geq 0,$$

since $r_\beta > 0$. We also get

$$\|\mathbf{x}_{\text{new}} - \mathbf{x}_{\text{old}}\| = \left\| \tilde{\mathbf{x}} + t\left(\mathbf{e}_v + \sum_{i=1}^m q_{i\beta}\mathbf{e}_{k_i} \right) - \tilde{\mathbf{x}} \right\| = t\left(1 + \sum_{i=1}^m q_{i\beta}^2 \right)^{1/2}.$$

This implies

$$\frac{\mathbf{c}^T(\mathbf{x}_{\text{new}} - \mathbf{x}_{\text{old}})}{\|\mathbf{x}_{\text{new}} - \mathbf{x}_{\text{old}}\|} = r_\beta \left/ \left(1 + \sum_{i=1}^m q_{i\beta}^2 \right)^{1/2} \right. .$$

In order to find the entering variable that maximizes this ratio, we have to know the entries of Q in all columns corresponding to indices β with $r_\beta > 0$. This requires a large number of extra computations in the usual implementation of the simplex method that does not explicitly maintain the tableau (see Section 5.6). According to Lemma 5.5.1, the computation of a single column of Q requires $O(m^2)$ arithmetic operations, and we may have to look at many of the $n - m$ columns.

The number of arithmetic operations can be brought down to $O(m)$ per column, by maintaining and updating the values

$$T_j = 1 + \sum_{i=1}^m q_{ij}^2, \quad j = 1, \ldots, n - m,$$

in every pivot step. If x_{k_α} is the leaving variable, the corresponding value in the next iteration is

$$T_j + \left(\frac{q_{\alpha j}}{q_{\alpha \beta}} \right)^2 T_\beta - 2\frac{q_{\alpha j}}{q_{\alpha \beta}} \sum_{i=1}^m q_{ij} q_{i\beta}, \tag{G.1}$$

and this value can be computed with $O(m)$ arithmetic operations for a single j, after a preprocessing that involves $O(m^2)$ arithmetic operations. The Devex pivot rule differs from this procedure as follows. First, it maintains a *reference framework* of only $n - m$ variables. For T_j, a value q_{ij}^2 is taken into account only if the variable x_{k_i} is in the reference framework. This has the effect that steepness of edges is measured only in the subspace spanned by the $n - m$ variables in the reference framework, and not in the space of all n variables. The major advantage of this approximation is that it is easy to set up: We have $T_j = 1$ for all j if the reference framework is initially chosen as the current set of nonbasic variables. Second, Devex maintains only an approximation $\tilde{T}_j$ of T_j, which is updated as follows: After each pivot step, $\tilde{T}_j$ is set to

$$\max \left(\tilde{T}_j, \left(\frac{q_{\alpha j}}{q_{\alpha \beta}} \right)^2 \tilde{T}_\beta \right).$$

This avoids the expensive computation of the sum in (G.1) and replaces the sum of the first two terms by their maximum. That way, the $\tilde{T}_j$ steadily grow and at some point they become bad approximations. This makes it necessary to reset the reference framework to the current set of nonbasic variables from time to time, meaning that all $\tilde{T}_j$ are reset to 1. Details and heuristic justifications for the simplified update formula are given in

> P. M. J. Harris: Pivot selection methods of the Devex LP code, *Math. Prog.* 5(1973) 1–28.

Dual simplex method. This is a method for solving linear programs in equational form. Like the simplex method, it goes through a sequence of simplex tableaus

$$\begin{array}{rcl} \mathbf{x}_B & = & \mathbf{p} + Q\mathbf{x}_N \\ \hline z & = & z_0 + \mathbf{r}^T \mathbf{x}_N \end{array}$$

until a tableau with $\mathbf{p} \geq \mathbf{0}$, $\mathbf{r} \leq \mathbf{0}$ is encountered. Such a tableau proves that the vector $\mathbf{x}^*$ given by $\mathbf{x}_B^* = \mathbf{p}, \mathbf{x}_N^* = \mathbf{0}$ is optimal. While the simplex method maintains the invariant $\mathbf{p} \geq \mathbf{0}$ (B is a feasible basis), the dual simplex method maintains the invariant $\mathbf{r} \leq \mathbf{0}$ (B is a *dual feasible* basis). As long as there are indices i such that $p_i < 0$, the dual simplex method chooses a leaving variable $x_u = x_{k_\alpha}$ with $p_\alpha < 0$. Then it searches for an entering variable $x_v = x_{\ell_\beta}$ whose increment results in $x_u = 0$. This is possible only if $q_{\alpha\beta} > 0$. Moreover, in the tableau corresponding to the next basis $B' = (B \setminus \{u\}) \cup \{v\}$, all coefficients of nonbasic variables in the last row should still be nonpositive. Rewriting the tableau as usual, we find that the next basis B' is dual feasible if and only if β satisfies

$$q_{\alpha\beta} > 0 \quad \text{and} \quad \frac{r_\beta}{q_{\alpha\beta}} = \max \left\{ \frac{r_j}{q_{\alpha j}} : q_{\alpha j} > 0, j = 1, 2, \ldots, n - m \right\}.$$

If the dual simplex method does not cycle, it will eventually reach $\mathbf{p} \geq \mathbf{0}$ (an optimal solution is found), or it encounters a situation in which all $q_{\alpha j}$ are nonpositive. But then the linear program under consideration is infeasible.

The reader might have noticed that the computations involved in a pivot step of the dual simplex method look very similar to those in the (primal) simplex method explained in Section 5.6. Indeed, they *are* the computations of the primal simplex method applied to the "dual" tableau

$$\begin{array}{rcl} \mathbf{y}_N & = & -\mathbf{r} - Q^T \mathbf{y}_B \\ \hline z & = & -z_0 - \mathbf{p}^T \mathbf{y}_B \end{array}$$

One can in fact show that the dual simplex method is just the primal simplex method in disguise, applied to the dual linear program, and so the dual simplex method is not really a new tool for solving linear programs. However, it is useful in practice, since it works with the original linear program (as opposed to the dual linear program), and most simplex-based linear programming solvers offer an option of using it. For certain classes of problems it can be substantially faster than the primal simplex method.

The dual simplex method is also useful after adding cutting planes, since the search for the new solution can start from the old one (which remains dual feasible).

Gomory cut. We consider a linear program

$$\text{maximize } \mathbf{c}^T \mathbf{x} \text{ subject to } A\mathbf{x} = \mathbf{b}, \ \mathbf{x} \geq \mathbf{0}$$

in equational form with $\mathbf{c} \in \mathbb{Z}^n$, along with an optimal basic feasible solution $\mathbf{x}^*$. A Gomory cut is a specific cutting plane for this linear program, derived from a feasible basis B associated with $\mathbf{x}^*$. Given B, we can rewrite the equations $A\mathbf{x} = \mathbf{b}$ and $z = \mathbf{c}^T \mathbf{x}$ in tableau form

$$\begin{array}{rcl} \mathbf{x}_B &=& \mathbf{p} + Q\mathbf{x}_N \\ \hline z &=& z_0 + \mathbf{r}^T \mathbf{x}_N \end{array}$$

(see Section 5.5). It is easy to show that if for some $i \in \{1, 2, \ldots, m\}$, the value $p_i = x_i^*$ is nonintegral, then the inequality

$$x_i \leq \lfloor p_i \rfloor + \lceil \mathbf{q}_i \rceil \mathbf{x}_N$$

is a cutting plane, where $\mathbf{q}_i$ is the ith row of Q. This cutting plane is called a Gomory cut. A special Gomory cut is obtained if $z_0 \notin \mathbb{Z}$:

$$z = \mathbf{c}^T \mathbf{x} \leq \lfloor z_0 \rfloor + \lceil \mathbf{r}^T \rceil \mathbf{x}_N.$$

We may now add the Gomory cuts as new inequalities to the linear program and recompute the optimal solution. Let us call this a *round*. The remarkable property of Gomory cuts is that we get an integral optimal solution after a finite number of rounds (assuming rational data). Since we never cut off integral solutions of the original linear program, this final solution is an optimal solution of the integer program

$$\text{maximize } \mathbf{c}^T \mathbf{x} \text{ subject to } A\mathbf{x} = \mathbf{b}, \ \mathbf{x} \geq \mathbf{0}, \ \mathbf{x} \in \mathbb{Z}^n.$$

The method of Gomory cuts is a simple but inefficient algorithm for solving integer programs.

Phases I and II. Traditionally, the computation of the simplex method is subdivided into *phase I* and *phase II*. In phase I, the auxiliary linear program for finding an initial feasible basis is solved, and phase II solves the original linear program, starting from this initial feasible basis (Section 5.6).

Phase transition. Even if phase I of the simplex method reveals that the original problem is feasible, it may happen that some of the auxiliary variables are still basic (with value 0) in the final feasible basis of the auxiliary problem. We have indicated in Section 5.6 that it is nevertheless easy to get a feasible basis for the original problem.

In the simplex method this part can elegantly be implemented by pivot steps of a special kind, in which the possible leftovers of the auxiliary variables x_{n+1} through x_{n+m} are forced to leave the basis. These pivot steps are said to form the *phase transition*.

Under our assumption that the matrix A has rank m, the phase transition is guaranteed to succeed. But even the case in which A has rank smaller than m can be handled during these special pivot steps, and this is an important aspect in practice, since it allows the simplex method to work with any input. If A does not have full rank, the consequence is that some of the auxiliary variables x_{n+1} through x_{n+m} cannot be forced to leave the basis. But any such variable can be shown to correspond to a redundant constraint, one that may be deleted from the linear program without changing the set of feasible solutions. Moreover, after deleting all the redundant constraints discovered in this way, the resulting subsystem $A'\mathbf{x} = \mathbf{b}'$ has a matrix A' of full rank. A basis of the linear program with respect to this new system is obtained from the current basis by removing the auxiliary variables that could not be forced to leave.

Pivot column. The column of the simplex tableau corresponding to the entering variable is called a pivot column. Depending on the context, this may or may not include the coefficient of the vector $\mathbf{r}$ corresponding to the entering variable.

Pivot element. This is the element of the simplex tableau in the pivot row and pivot column.

Pivot row. The row of the simplex tableau corresponding to the leaving variable is called the pivot row. Depending on the context, this may or may not include the coefficient of the vector $\mathbf{p}$ that holds the value of the leaving variable.

Pricing. The process of selecting the entering variable during a pivot step of the simplex method is sometimes referred to as pricing. We say that a nonbasic variable x_{ℓ_j} is *priced* when its coefficient r_j in the last row of the simplex tableau is computed (Section 5.6).

Primal–dual method. This is a method for solving a linear program by iteratively improving a feasible solution of the dual linear program. Let us start with a linear program in equational form:

$$\text{Maximize } \mathbf{c}^T\mathbf{x} \text{ subject to } A\mathbf{x} = \mathbf{b} \text{ and } \mathbf{x} \geq \mathbf{0}. \tag{P}$$

We assume that $\mathbf{b} \geq \mathbf{0}$ and that (P) is bounded. The dual linear program is

$$\text{minimize } \mathbf{b}^T\mathbf{y} \text{ subject to } A^T\mathbf{y} \geq \mathbf{c}. \tag{D}$$

Let us assume that we have a feasible solution $\tilde{\mathbf{y}}$ of (D). Then we define

$$J = \{j \in \{1, 2, \ldots, n\} : \mathbf{a}_j^T \mathbf{y} = c_j\},$$

where $\mathbf{a}_j$ denotes the jth column of A. It turns out that $\tilde{\mathbf{y}}$ is an optimal solution of (D) if and only if there exists a feasible solution $\tilde{\mathbf{x}}$ of (P) such that

$$\tilde{x}_j = 0 \text{ for all } j \in \{1, 2, \ldots, n\} \setminus J \tag{G.2}$$

(this can easily be checked directly or seen from the complementary slackness theorem). To check for the existence of such a vector $\tilde{\mathbf{x}}$, we solve an auxiliary linear program, the *restricted primal*:

$$\begin{aligned} \text{Minimize} \quad & z_1 + z_2 + \cdots + z_m \\ \text{subject to} \quad & A_J \mathbf{x}_J + I_m \mathbf{z} = \mathbf{b} \\ & \mathbf{x}, \mathbf{z} \geq \mathbf{0}. \end{aligned} \tag{RP}$$

By our assumption $\mathbf{b} \geq \mathbf{0}$, (RP) is feasible, and it is also bounded. Let $\mathbf{x}^*, \mathbf{z}^*$ be an optimal solution. If $\mathbf{z}^* = \mathbf{0}$, then $\mathbf{x}^*$ optimally solves (P), and $\tilde{\mathbf{y}}$ optimally solves (D). Otherwise, we know that $\tilde{\mathbf{y}}$ cannot be an optimal solution of (D), and we want to find a better dual solution. To this end, we consider the dual of (RP), which can be written as

$$\begin{aligned} \text{maximize} \quad & \mathbf{b}^T \mathbf{y} \\ \text{subject to} \quad & (A_J)^T \mathbf{y} \leq \mathbf{0} \\ & \mathbf{y} \leq \mathbf{1}. \end{aligned} \tag{RD}$$

Let $\mathbf{y}^*$ be an optimal solution of (RD). From the duality theorem we know that $\mathbf{b}^T \mathbf{y}^* = \sum_{i=1}^m z_i^* > 0$. Consequently, for every $t > 0$, the vector $\tilde{\mathbf{y}} - t\mathbf{y}^*$ is an improved dual solution, provided that it is feasible for (D). We claim that there exists a small $t > 0$ such that $\tilde{\mathbf{y}} - t\mathbf{y}^*$ actually is feasible for (D). Indeed, for $j \in J$, we have

$$\mathbf{a}_j^T (\tilde{\mathbf{y}} - t\mathbf{y}^*) \geq c_j + t \cdot 0,$$

and for $j \in \{1, 2, \ldots, n\} \setminus J$, we get

$$\mathbf{a}_j^T (\tilde{\mathbf{y}} - t\mathbf{y}^*) > c_j - t\mathbf{a}_j^T \mathbf{y}^* \geq c_j$$

for a suitable t. Now we choose t^* as large as possible such that the last inequality still holds for all $j \in \{1, 2, \ldots, n\} \setminus J$, and we replace $\tilde{\mathbf{y}}$ by $\tilde{\mathbf{y}} - t^*\mathbf{y}^*$ for the next iteration of the primal–dual method. The set J will change as well, since at least one inequality that previously had slack has now become tight. Note that t^* exists, since otherwise, (D) is unbounded and (P) is infeasible, in contradiction to our assumption $\mathbf{b} \geq \mathbf{0}$.

With some care, this method yields an optimal solution of (D), and hence an optimal solution $\mathbf{x}^*$ to (RP) and (P), after a finite number of iterations.

This is not a priori clear: Even if the dual objective function strictly decreases in every iteration, the improvement could become smaller and smaller; in fact, there are examples in which this happens, and in which sets J reappear infinitely often. But this can be overcome by choosing for example the lexicographically largest optimal solution $\mathbf{y}^*$ of (RD).

There are two aspects that make the primal–dual method attractive. First of all, the restricted primal (RP) and its dual (RD) do not involve the original objective function vector $\mathbf{c}$. This means that the primal–dual method reduces an optimization problem to a sequence of decision problems. These decision problems are in many cases much easier to solve and do not require linear programming techniques. For instance, if (P) is the linear program that computes a maximum-weight matching in a bipartite graph (Section 3.2), then (RP) computes a matching of maximum cardinality in the subgraph with edges indexed by J. Moreover, if (RP) has been solved once, its solution in the next iteration is easy to get through an augmenting path. These insights form the basis of the *Hungarian method* for maximum-weight matchings, which can be interpreted as a sophisticated implementation of the primal–dual method. A second aspect appears in connection with approximation algorithms for NP-hard problems. Starting from the LP relaxation (P) of a suitable integer program, we may search for a vector $\tilde{\mathbf{x}}$ that satisfies a condition weaker than (G.2), but in return it allows us to construct a related integral feasible solution $\mathbf{x}^*$ to (P). If we cannot find $\tilde{\mathbf{x}}$, we know as before that the dual solution $\tilde{\mathbf{y}}$ can be improved. If we find $\tilde{\mathbf{x}}$, then $\tilde{\mathbf{y}}$ may not yet be optimal, but we may still be able to argue that $\mathbf{x}^*$ is a reasonably good solution of the integer program. A number of approximation algorithms based on the primal–dual method are described in

> M. X. Goemans and D .P. Williamson: The primal–dual method for approximation algorithms and its application in network design problems, in *Approximation Algorithms* (D. Hochbaum, editor), PWS Publishing Company, Boston, 1997, pages 144–191.

Ratio test.   The process of selecting the leaving variable during a pivot step of the simplex method is called the ratio test. The leaving variable x_{k_α} is such that it has *minimum ratio* $-\frac{p_\alpha}{q_{\alpha\beta}}$ (Section 5.6).

Reduced costs.   The vector $\mathbf{r}$ in a simplex tableau; r_j is the reduced cost of variable x_{ℓ_j} (Section 5.5).

Sensitivity analysis.   The components of the matrix A and the vectors $\mathbf{b}$ and $\mathbf{c}$ that define a linear program are often results of measurements or estimates. Then an important question is how "stable" the optimal solution $\mathbf{x}^*$ is. Ideally, if the components vary by small amounts, then $\mathbf{x}^*$ (or at least $\mathbf{c}^T \mathbf{x}^*$) varies by a small amount as well. In this case small errors in collecting data can safely be ignored.

It may well be that small changes in some of the components have a more drastic impact on $\mathbf{x}^*$. Sensitivity analysis tries to assess how the solution depends on the input, both by theoretical considerations and by computer simulations.

The simplex method and the dual simplex method are excellent tools for the latter task, given that only $\mathbf{b}$ and $\mathbf{c}$ vary. If $\mathbf{c}$ varies, we can start from the old optimal solution $\mathbf{x}^*$ and reoptimize with respect to the new objective function vector $\mathbf{c}$. If $\mathbf{c}$ changed only a little, chances are good that this reoptimization requires only few pivot steps.

Similarly, if $\mathbf{b}$ changes, the dual simplex method may be used to reoptimize, since the old optimal solution $\mathbf{x}^*$ is still associated with a dual feasible basis under the new right-hand-side vector $\mathbf{b}$.

The primal simplex method can also efficiently deal with the addition of new variables, while the dual simplex method can handle the addition of constraints. These two operations in particular occur in connection with column generation and cutting planes.

Total dual integrality. A system $A\mathbf{x} \leq \mathbf{b}$ of linear inequalities is said to be totally dual integral (TDI) if the linear program

$$\text{minimize } \mathbf{b}^T\mathbf{y} \text{ subject to } A^T\mathbf{y} = \mathbf{c}, \ \mathbf{y} \geq \mathbf{0}$$

has an integral optimal solution $\tilde{\mathbf{y}}$ whenever $\mathbf{c}$ is an *integral* vector for which an optimal solution exists. It can then be shown that the (primal) problem

$$\text{maximize } \mathbf{c}^T\mathbf{x} \text{ subject to } A\mathbf{x} \leq \mathbf{b}$$

has an integral optimal solution for *all* $\mathbf{c}$, provided that A is rational and $\mathbf{b}$ is integral.

Under total dual integrality of $A\mathbf{x} \leq \mathbf{b}$, the set $P = \{\mathbf{x} : A\mathbf{x} \leq \mathbf{b}\}$ is therefore an integral polyhedron (provided A is rational and $\mathbf{b}$ is integral). We first met integral polyhedra in Section 8.2, where we used the concept of total unimodularity to establish integrality. The TDI notion is connected to total unimodularity as follows. A matrix A is totally unimodular if and only if the system $\{A\mathbf{x} \leq \mathbf{b}, \mathbf{x} \geq \mathbf{0}\}$ is TDI for all integral vectors $\mathbf{b}$.

One of the most prominent examples of a TDI system is Edmonds' description of the **matching polytope** by inequalities. For a (not necessarily bipartite) graph $G = (V, E)$, the matching polytope is the convex hull of all *incidence vectors* of matchings. As in Section 3.2, the incidence vector of a matching $M \subseteq E$ is the $|E|$-dimensional vector $\mathbf{x}$ with

$$x_e = \begin{cases} 1 & \text{if } e \in M, \\ 0 & \text{otherwise.} \end{cases}$$

Any such vector satisfies the inequalities

$$\begin{aligned} x_e &\geq 0, & e \in E, \\ \sum_{e \ni v} x_e &\leq 1, & v \in V. \end{aligned} \tag{G.3}$$

If G is a bipartite graph, we have shown that the polytope P defined by these inequalities is integral (Section 8.2, Lemmas 8.2.5 and 8.2.4). It follows that every vertex of P is the incidence vector of some matching. Consequently, P coincides with the matching polytope in the bipartite case, and the system (G.3) is an explicit description of the matching polytope by inequalities.

This is no longer true for nonbipartite graphs. Indeed, if we consider the triangle

then the polytope P defined by (G.3) has the five vertices $\mathbf{0}$, $\mathbf{e}_1$, $\mathbf{e}_2$, $\mathbf{e}_3$, and $\frac{1}{2}(\mathbf{e}_1 + \mathbf{e}_2 + \mathbf{e}_3)$ and is therefore not integral:

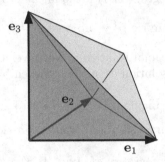

Yet there is a larger inequality system that leads to an integral polytope. The additional inequalities stem from the observation that every subset $A \subseteq V$ of odd size $2k+1$ supports at most k edges of any fixed matching. Therefore, every incidence vector of a matching satisfies all inequalities in the following system (the *odd subset inequalities*):

$$\sum_{e \subseteq A} x_e \leq \frac{|A| - 1}{2}, \qquad \text{for all } A \subseteq E \text{ with } |A| \text{ odd.} \qquad \text{(G.4)}$$

It can be shown that the system of inequalities in (G.3) and (G.4) is TDI, and consequently, these inequalities define the matching polytope for a general graph G.

In the case of a triangle, the only nontrivial inequality in (G.4) is obtained for $A = V$:

$$\sum_{e \in E} x_e \leq 1.$$

This inequality cuts off the fractional vertex $\frac{1}{2}(\mathbf{e}_1 + \mathbf{e}_2 + \mathbf{e}_3)$ and leaves the tetrahedron that is the convex hull of $\mathbf{0}$, $\mathbf{e}_1$, $\mathbf{e}_2$, and $\mathbf{e}_3$. This integral polytope is the matching polytope of the triangle.

We now have an exponentially large but explicit inequality description of the matching polytope of a general graph. Having such a description for an integral polyhedron is an indication that it might be possible to optimize a linear function over the polyhedron in polynomial time. In the case of matchings in general graphs, this can indeed be done. A maximum-weight matching in a general graph can be found in polynomial time, but the known algorithms are much more involved than those for the bipartite case discussed in Section 3.2.

Index

Universitext

Holmgren, R. A.: A First Course in Discrete Dynamical Systems

Howe, R., Tan, E. Ch.: Non-Abelian Harmonic Analysis

Howes, N. R.: Modern Analysis and Topology

Hsieh, P.-F.; Sibuya, Y. (Eds.): Basic Theory of Ordinary Differential Equations

Humi, M., Miller, W.: Second Course in Ordinary Differential Equations for Scientists and Engineers

Hurwitz, A.; Kritikos, N.: Lectures on Number Theory

Huybrechts, D.: Complex Geometry: An Introduction

Isaev, A.: Introduction to Mathematical Methods in Bioinformatics

Istas, J.: Mathematical Modeling for the Life Sciences

Iversen, B.: Cohomology of Sheaves

Jacod, J.; Protter, P.: Probability Essentials

Jennings, G. A.: Modern Geometry with Applications

Jones, A.; Morris, S. A.; Pearson, K. R.: Abstract Algebra and Famous Impossibilities

Jost, J.: Compact Riemann Surfaces

Jost, J.: Dynamical Systems. Examples of Complex Behaviour

Jost, J.: Postmodern Analysis

Jost, J.: Riemannian Geometry and Geometric Analysis

Kac, V.; Cheung, P.: Quantum Calculus

Kannan, R.; Krueger, C. K.: Advanced Analysis on the Real Line

Kelly, P.; Matthews, G.: The Non-Euclidean Hyperbolic Plane

Kempf, G.: Complex Abelian Varieties and Theta Functions

Kitchens, B. P.: Symbolic Dynamics

Kloeden, P.; Ombach, J.; Cyganowski, S.: From Elementary Probability to Stochastic Differential Equations with MAPLE

Kloeden, P. E.; Platen; E.; Schurz, H.: Numerical Solution of SDE Through Computer Experiments

Kostrikin, A. I.: Introduction to Algebra

Krasnoselskii, M. A.; Pokrovskii, A. V.: Systems with Hysteresis

Kurzweil, H.; Stellmacher, B.: The Theory of Finite Groups. An Introduction

Lang, S.: Introduction to Differentiable Manifolds

Luecking, D. H., Rubel, L. A.: Complex Analysis. A Functional Analysis Approach

Ma, Zhi-Ming; Roeckner, M.: Introduction to the Theory of (non-symmetric) Dirichlet Forms

Mac Lane, S.; Moerdijk, I.: Sheaves in Geometry and Logic

Marcus, D. A.: Number Fields

Martinez, A.: An Introduction to Semiclassical and Microlocal Analysis

Matoušek, J.: Using the Borsuk-Ulam Theorem

Matsuki, K.: Introduction to the Mori Program

Mazzola, G.; Milmeister G.; Weissman J.: Comprehensive Mathematics for Computer Scientists 1

Mazzola, G.; Milmeister G.; Weissman J.: Comprehensive Mathematics for Computer Scientists 2

Mc Carthy, P. J.: Introduction to Arithmetical Functions

McCrimmon, K.: A Taste of Jordan Algebras

Meyer, R. M.: Essential Mathematics for Applied Field

Meyer-Nieberg, P.: Banach Lattices

Mikosch, T.: Non-Life Insurance Mathematics

Mines, R.; Richman, F.; Ruitenburg, W.: A Course in Constructive Algebra

Moise, E. E.: Introductory Problem Courses in Analysis and Topology